Bisociative Literature-Based Discovery

Nada Lavrač
Bojan Cestnik
Andrej Kastrin

Bisociative Literature-Based Discovery

Methods with Tutorials in Python

Springer

Nada Lavrač
Jožef Stefan Institute
Ljubljana, Slovenia

Bojan Cestnik
Temida d.o.o.
Ljubljana, Slovenia

Andrej Kastrin
Faculty of Medicine
University of Ljubljana
Ljubljana, Slovenia

ISBN 978-3-031-96862-4 ISBN 978-3-031-96863-1 (eBook)
https://doi.org/10.1007/978-3-031-96863-1

This Springer imprint is published by the registered company Springer Nature Switzerland AG
The registered company address is: Gewerbestrasse 11, 6330 Cham, Switzerland

This monograph is dedicated to our families.

Foreword

I fondly remember research project BISON[1] where Nada's and my team—together with a number of colleagues from all over Europe—explored various aspects of bisociative knowledge discovery. At its heart, the project aimed at supporting humans in their quest to find new, creative insights into our world by exploiting a much broader set of information repositories. As opposed to searching for frequent patterns and/or bridges in closely related domains, bisociative discovery explores also rare cross-domain associations (termed "bisociations") among vaguely or completely unrelated domains, triggering novel, often serendipitous insights. Methods based on network analysis seemed most powerful at that time, and domains represented in textual form the most promising information source to find these types of connections.

After BISON ended, many of the members of the consortium continued to work on various aspects of bisociative knowledge discovery. The book by Nada, Bojan, and Andrej summarizes lots of work from Nada and her colleagues around literature-based discovery. They explore different types of textual representations and address the key aspects of bisociative knowledge discovery: finding (interesting) outliers (supporting the famous "uh, this is weird" eureka-moments), identification of bridging concepts (triggering cross-domain discovery, e.g. "if this works here, what would this look like over here?"), and network-based methods (supporting discoveries of similar dependencies or interactions across domains). In addition to those closed discovery approaches, they also discuss methods for open discovery. The book is rounded off by an extensive state-of-the-art introduction and many practical examples, allowing us to base own experiments on the presented material. It makes for a great introduction to the field but also lends itself to browse around and enjoy the many nuggets of information sprinkled all over the volume.

Although in our field predictions about the future are feared, I dare to predict that methods as described in this book will have increasing impact on the field of true knowledge discovery, even with the current rise of GenAI. Purely data driven discovery

[1] Project BISON (Bisociation Networks for Creative Information Discovery) was funded under the European Commission's Seventh Framework Programme for Research and Technological Development (FP7).

resembles a poking around in the fog, ultimately finding only consensus. More focused and (sometimes: complex) structure-driven approaches allow us to find truly new insights in large and highly diverse information repositories.

Konstanz, September 2024

Michael Berthold

Preface

Literature-based discovery (LBD) is a process of uncovering new knowledge by analyzing and connecting disparate pieces of information from different sources of literature. Its primary goal is to generate novel hypotheses or insights that are not apparent from the domain-specific documents of the expert's primary interest.

This monograph, titled *Bisociative Literature-Based Discovery*, establishes an integrated view on LBD grounded in two research traditions: bisociative knowledge discovery (Berthold, 2012) and literature-based discovery (Swanson, 1986). By combining these traditions, we emphasize the potential of bisociative reasoning—a creative process involving the connection of seemingly unrelated concepts (Koestler, 1964)—to support the formation of serendipitous hypotheses.

The monograph introduces the field of bisociative LBD by first explaining the underlying LBD principles and techniques, followed by the presentation of bisociative LBD techniques and applications developed by the authors. Selected techniques include conventional natural language processing (NLP) approaches, as well as outlier-based, concept-based, network-based, and embeddings-based LBD approaches. Reproducibility aspects of bisociative LBD research are also covered, addressing all steps of the bisociative LBD process: data acquisition, text preprocessing, hypothesis discovery, and evaluation.

The extensive coverage of bisociative LBD offers numerous benefits to the reader.

1. Improved understanding of the elementary components of LBD methods. Systematic coverage of basic LBD concepts and approaches, through a brief survey of the LBD field (Chapter 2), and an overview of related background technologies (Chapter 3), offer the readers a gentle introduction to LBD.
2. Hands-on LBD experience and replicability. An open science approach (Chapter 4) supports experiment replicability and software reuse: the last sections of most of the chapters contain tutorials illustrating how to implement simplified LBD workflows using the popular Jupyter Notebooks software.
3. Enhanced text analysis with cross-domain insights. Using traditional Bag-of-Words (BoW) representations of text documents, bisociative discoveries are facilitated by an ensemble-based approach to identifying cross-domain bridging terms with the CrossBee LBD system (Chapter 5).

4. Targeted outlier-based discovery. Outlier-based bisociative LBD methods efficiently narrow down the search for cross-domain bridging terms to outlier documents, simplifying the exploration of bridging terms by domain experts (Chapter 6, and Section 7.3 of Chapter 7).
5. Leveraging domain knowledge with concept-based approaches. By exploiting domain knowledge captured in taxonomies, ontologies, or concept maps, utilizing public resources such as Unified Medical Language System (UMLS) concepts within the Medical Subject Headings (MeSH) and Gene Ontology (GO), concept-based methods facilitate cross-domain knowledge exploration (Section 7.2 of Chapter 7).
6. Uncovering relationships with network-based approaches. Network-based approaches extend bridging term discovery to relation discovery, exploiting their potential to reveal bisociative relationships among concepts e.g., across different biomedical domains (Chapter 8).
7. Advanced data processing via embeddings. Embeddings-based data processing techniques enable the extraction of semantic similarities between concepts and relations in a unified vector representation space, which greatly enhances cross-domain LBD (Chapter 9).
8. Literature analysis with large language models (LLMs). We also explore an LLM-based approach to structuring LBD literature, offering novel domain insights (Chapter 10).

The monograph is targeted at researchers, students, and domain experts interested in bisociative knowledge exploration from large public corpora, such as MEDLINE/PubMed. By covering texts, relations, networks, and ontologies, this work empowers domain experts to transcend their knowledge silos when confronted with varied data formats in their research practice. The monograph's open science approach with tutorials in Python allows for code reuse and experiment replicability. These tutorials and related Jupyter notebooks are accessible through the book's GitHub repository at https://github.com/lbd-book/code.git.

We acknowledge numerous collaborators who contributed to our past LBD research. These include data scientists Tanja Urbančič, Blaž Škrlj, Borut Sluban, Senja Pollak, Ingrid Petrič, Matic Perovšek, Matej Martinc, Boshko Koloski, Matjaž Juršič, Dimitar Hristovski, and domain experts Maruša Pompe, Marta Macedoni Lukšič, Elsa Fabretti. We are grateful to Lenka Trdina for the graphical design of figures, and to Ljupčo Todoroski for the comments on the draft monograph text. We are also grateful to Michael Berthold for providing an insightful foreword to this monograph.

We acknowledge our institutions, Jožef Stefan Institute, University of Ljubljana, and University of Nova Gorica, Slovenia, for providing stimulating working environments for our research. We acknowledge the financial support of the Slovenian Research and Innovation Agency core research programmes P2-0103 and P3-0154 for co-funding our research. We are also grateful to Springer Nature for their support in publishing this monograph.

Ljubljana,
December 2024

Nada Lavrač
Bojan Cestnik
Andrej Kastrin

Contents

1 **Introduction** ... 1
 1.1 Motivation ... 1
 1.2 Literature-based discovery at a glance ... 3
 1.3 Underlying principles and theories ... 7
  1.3.1 Design principles of creativity support tools ... 7
  1.3.2 Concept creation and discovery ... 8
  1.3.3 Transformative discovery ... 9
  1.3.4 Bisociative reasoning ... 10
  1.3.5 Types of bisociations ... 10
  1.3.6 Relating Swanson's ABC model and bisociative LBD ... 12
 1.4 Evaluation of LBD approaches ... 13
  1.4.1 Discovery replication ... 14
  1.4.2 Time-based evaluation ... 14
 1.5 Outline of the monograph and the approach to code reuse ... 15

2 **History, Resources and Tools** ... 21
 2.1 Historical sketch ... 21
 2.2 Summary of LBD review papers ... 25
 2.3 LBD research topics and keywords ... 26
  2.3.1 Word cloud visualization of the field ... 27
  2.3.2 Analysis of topics through a scientometric lens ... 27
  2.3.3 Longitudinal keywords analysis ... 28
  2.3.4 Longitudinal analysis of two selected topics ... 29
 2.4 Standard LBD resources ... 31
  2.4.1 Public literature resources ... 31
  2.4.2 Semantic LBD resources ... 32
 2.5 Selected LBD tools ... 36
  2.5.1 Selected document acquisition tools ... 36
  2.5.2 Selected LBD support tools ... 38

3 Background Technologies 45
3.1 Information extraction 45
3.2 Natural language processing 46
3.3 Text mining 48
3.3.1 Bag of Words vector representations for text units 48
3.3.2 Similarity measures for text mining 49
3.4 Network analysis 50
3.4.1 Network representations 51
3.4.2 Network analysis tasks 52
3.5 Embeddings 53
3.6 Large Language Models 55
3.7 Evaluation of LBD approaches 56
3.7.1 Evaluation measures used in classification 57
3.7.2 The ROC space model 58
3.7.3 Evaluation measures used in link prediction 60
3.7.4 Qualitative evaluation of LBD support systems 61

4 Benchmark Data and Reusable Python Code 67
4.1 Benchmark data sets used for experiment replicability 67
4.2 Reusable processing components and LBD pipelines 70
4.2.1 Jupyter Python notebooks: Brief installation guidelines 71
4.2.2 Running Jupyter Notebooks using Docker 72
4.2.3 LBD Jupyter Python notebook implementations 72

5 Text Mining for Closed Discovery 77
5.1 Undiscovered public knowledge: A tribute to Don R. Swanson 77
5.1.1 LBD through serendipity 77
5.1.2 Arrowsmith LBD tool 80
5.1.3 Replicating the early discovery by Swanson 80
5.2 CrossBee methodology for bridging term detection 82
5.2.1 Heuristics for bridging term discovery 82
5.2.2 Ensemble heuristics 86
5.2.3 CrossBee's HCI interface for bridging term discovery 88
5.3 Tutorials in Python 90
5.3.1 01_closed_discovery.ipynb 90
5.3.2 03_mini_crossbee.ipynb 91

6 Outlier-based Closed Discovery 93
6.1 Motivation for using outliers in LBD 93
6.2 Document classification for outlier detection and b-term identification 94
6.3 Document clustering for outlier detection and b-term identification 96
6.4 Relating outlier detection to CrossBee heuristics 98
6.5 Methodology combining outlier detection and CrossBee document exploration 99
6.6 Application: Finding connections between Alzheimer's disease and gut microbiota 104

6.7 Discussion 106
6.8 Tutorial in Python 107
6.8.1 04_mini_ontogen.ipynb 107

7 Semantic and Outlier-based Open Discovery 111
7.1 Comparison of open and closed LBD 111
7.2 Concept-based approach to open discovery 113
7.2.1 Replicating Raynaud's disease–fish oil open discovery through concept mapping 113
7.2.2 Knowledge-based approaches to reduce the concept space 114
7.3 RaJoLink open discovery methodology 116
7.3.1 Semantic filtering in RaJoLink 117
7.3.2 RaJoLink methodology 118
7.3.3 Application: Finding connections between autism and calcineurin 121
7.3.4 Discussion 123
7.4 Tutorials in Python 124
7.4.1 02_open_discovery 125
7.4.2 05_mini_rajolink.ipynb 125

8 Network-based Closed Discovery 129
8.1 Co-occurrence and link prediction background 129
8.2 Link prediction approach to LBD 130
8.2.1 Experimental setup 131
8.2.2 Link prediction methodology 132
8.2.3 Performance evaluation of link prediction 133
8.2.4 Application: Finding connections between schizophrenia and inflammation 136
8.3 Discussion 137
8.4 Tutorial in Python 138
8.4.1 06_link_prediction.ipynb 138

9 Embedding-based Closed Discovery 141
9.1 Embeddings 141
9.2 Text embedding-based bisociative LBD methodology 143
9.2.1 Methodology of bridging by relational bisociation 144
9.2.2 Application: Finding connections between circadian rhythm and plant defense 145
9.2.3 Results evaluation 146
9.3 Network embedding-based bisociative LBD methodology 149
9.3.1 Knowledge graph preprocessing 150
9.3.2 Knowledge graph completion 151
9.4 Tutorials in Python 153
9.4.1 07_find_genes.ipynb 153
9.4.2 08_drug_repurposing.ipynb 154

10 Research Trends and Lessons Learned 157
10.1 Selected lessons learned from past LBD research 157
10.2 Towards LBD using Large Language Models 161
10.3 Using LLMs for LBD literature structuring 162
10.3.1 Domain-adaptation via sentence-transformers 163
10.3.2 Prompt engineering of LLMs to design topic names 164
10.3.3 BERTopic modeling 164
10.3.4 The AHAM heuristic 165
10.3.5 Qualitative evaluation and future directions 166
10.4 Tutorial in Python 166
10.4.1 09_aham_topic_modeling 166

Index 171

Chapter 1
Introduction

Literature-based discovery (LBD) is a vibrant area of research, with a long research tradition. This chapter starts with a motivation for LBD research in Section 1.1, a brief introduction to LBD in Section 1.2, followed by an overview of the main underlying principles and related theories in Section 1.3. Finally, Section 1.5 offers a categorization of LBD tasks and approaches and provides an outline of this monograph.

1.1 Motivation

Increasing amounts of available knowledge and data exceed the analytical capabilities of humans. Take the corpus of biomedical papers in online bibliographical repositories, often referred to as the *bibliome*, which is of considerable size and complexity. For example, the PubMed database[1] comprises more than 36 million bibliographic records of life science literature, with a continuous influx of new records.[2] While the amount of biomedical literature available is growing at a very high rate, large chunks of potentially useful information in it remain unexploited (Cheng et al., 2014).

It is hard if not impossible for researchers to profit from all this information, even in their relatively narrow fields of research. In alignment with the theory of transformative discoveries[3], the growth of scientific knowledge strongly depends on the accumulation and synthesis of information from existing knowledge. Due to the huge amount of data, it is a difficult task to combine scientific information from different sources into coherently interpretable knowledge. Therefore, new technologies need to be developed and used to help analyze and extract useful information from large data and knowledge sources. In this context, the development of automated text mining and information extraction tools is of utmost importance.

Understanding complex phenomena and solving difficult problems often require combining knowledge from different domains and considering cross-domain associa-

[1] https://datadiscovery.nlm.nih.gov/

[2] A daily average is 2000–4000 of new PubMed records.

[3] See Section 1.3.3.

N. Lavrač et al., *Bisociative Literature-Based Discovery*,
https://doi.org/10.1007/978-3-031-96863-1_1

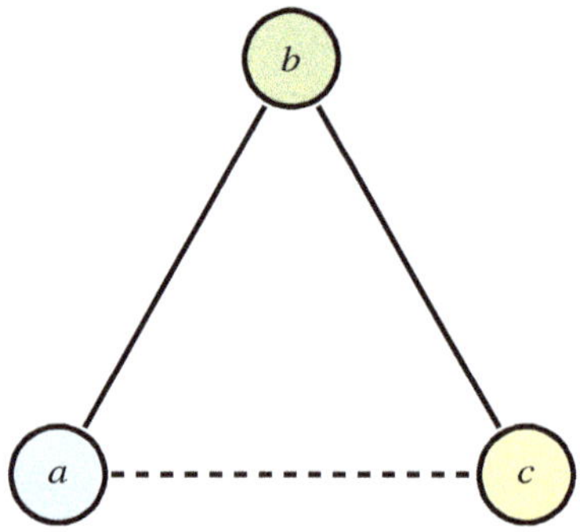

Fig. 1.1 Swanson's ABC model contains three concepts: the start concept (c), the intermediate concept (b), and the target concept (a). The LBD process begins with retrieving $c-b$ and $b-a$ relationships. Next, it combines associations with the same intermediates. Finally, it retrieves a list of $a-c$ relationships. If there is no prior mention of a particular $a-c$ connection, a hypothesis of a potential novel relationship between a and c concepts is conceived.

tions. The concept of association is at the heart of several information technologies, including information retrieval and data mining, while scientific discovery requires creative reasoning, including new insights that connect seemingly unrelated information. For example, using analogies or metaphors that connect concepts from different domains can lead to new insights into a given domain of interest. According to Koestler (1964), cross-context associations, referred to as *bisociations* in his seminal work *The Art of Creation*, are often necessary for innovative discoveries.

Numerous bisociative discovery methods, used in the search for cross-domain connections leading to potential new scientific discoveries (Dubitzky et al., 2012), employ the original text representation, or derived graph representations. It has been shown by Swanson (1990) that the discovery of bisociative knowledge can be addressed using text mining from the scientific literature, which is in this work referred to as LBD (Bruza & Weeber, 2008). In his early work, Swanson (1990) proposed a novel method to generate hypotheses across previously disjoint literature sets (separate document corpora A and C, related to concepts of interest a and c, respectively).[4] To find out whether a concept (term c) was associated with some other concept (term a)—although there was no direct evidence for this in the literature—Swanson searched for intermediate terms b representing potentially interesting concepts connected with c the articles of domain C, and with a in the articles of domain A. Putting these connections together and looking at their meaning proved to have the potential to provide new insights into the relationship between concepts c and a. This approach, referred to as Swanson's *ABC model*, is illustrated in Figure 1.1.

Let us explain this setting with an example that has become well-known in LBD. In his study, Swanson (1988, 1990) investigated whether magnesium deficiency could

[4] A *concept* can refer to any phenomenon, abstract notion, object, or idea. In the computational creativity literature (e.g., Boden, 1990), a key requirement of any computational model that manipulates concepts concerns the *representation* of a concept being manipulated. To simplify the matter, throughout this monograph, we use notions *concept* and *concept representation* interchangeably, while using linguistic *terms* to represent concepts.

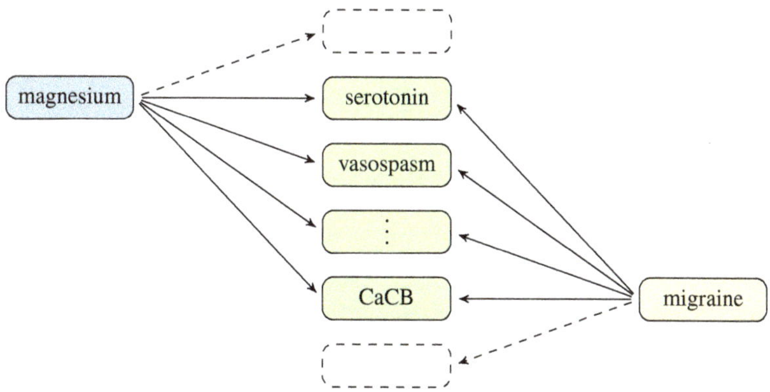

Fig. 1.2 Closed discovery when exploring migraine and magnesium documents, with *b*-terms identified by Swanson et al. (2006).

cause migraine. He found more than 60 pairs of articles—consisting of one article from the literature about migraine (c) and one article from the literature about magnesium (a)—connecting a with c through various third terms (b). For example, in the literature about magnesium, he found a statement that magnesium is a natural calcium channel blocker, while in the literature about migraine (intense headache), he discovered a statement that calcium channel blockers can prevent migraine attacks. In this case, calcium channel blockers act as a *bridge* between the concepts magnesium (a) and migraine (c). A closer inspection showed that eleven pairs of statements identified in different documents were suggestive and supportive of the hypothesis that magnesium deficiency may cause migraine attacks (Swanson, 1990), one of them including the term calcium channel blocker, as shown in Figure 1.2.

The initial work of Swanson relied on manual literature inspection, but many researchers followed this idea and developed the means of automatically or semi-automatically searching for *bridging terms*, linking separate domains through literature mining. An overview of the literature-based discovery approaches and challenges is provided by Bruza and Weeber (2008). Govaert and Nadif (2014) demonstrated that some other data mining approaches, such as co-clustering, may be suitable for identifying certain types of bisociations.

1.2 Literature-based discovery at a glance

LBD is the process of uncovering new knowledge by analyzing and connecting disparate pieces of information across existing literature sources. Its primary goal is to generate novel hypotheses or insights that might not be revealed from domain-specific documents of the expert's primary interest.

In cross-domain LBD, which addresses knowledge discovery from two (or more) initially separate document corpora, a crucial LBD step is the identification of interesting *bridging terms* (b-terms) or *bridging links* (b-links) that carry the potential of explicitly

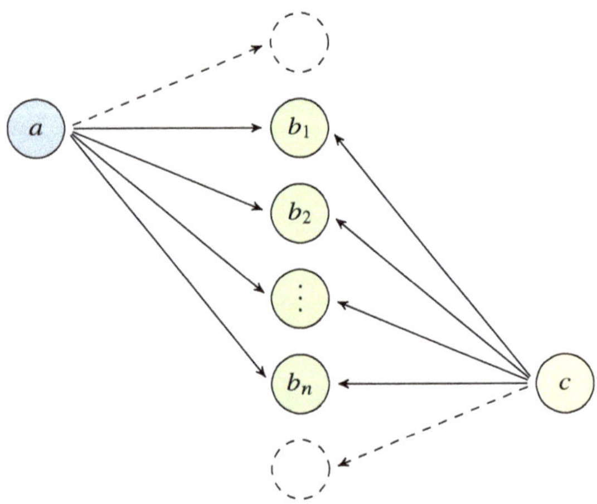

Fig. 1.3 Closed discovery process defined by Weeber et al. (2001), where the goal is to discover bridging concepts $b_1, b_2, \ldots, b_n$.

revealing the connections between the separate domains. Early LBD approaches (Smalheiser & Swanson, 1998; Swanson, 1990) focused on detecting interesting b-terms to uncover possible cross-domain relations between previously unrelated concepts. For example, the Arrowsmith online system (Smalheiser & Swanson, 1998) took as input two sets of titles of scientific papers from disjoint domains (disjoint document corpora) A and C, and listed terms that are common to A and C; the resulting bridging terms (b-terms) were further investigated by the user for their potential to generate new scientific hypotheses[5]. This approach is aligned with *ABC model of knowledge discovery*. The approach complies with the so-called closed discovery setting defined as follows.

Closed discovery. The closed discovery setting proposed by Weeber et al. (2001), where two initially separate domains A and C are specified by the user at the beginning of the discovery process, and the goal is to search for bridging concepts (terms) b in B to validate the hypothesized connection between A and C. The closed discovery setting, which is the most frequently addressed LBD setting, is illustrated in Figure 1.3.

Swanson's influential work has shown that databases such as PubMed can serve as a rich source of yet hidden relations between usually unrelated topics, potentially leading to novel insights and discoveries. By studying two separate pieces of literature—the literature on migraine headache and the literature on magnesium—Swanson (1988) discovered *eleven neglected connections*, all of them supportive of the hypothesis that magnesium deficiency might cause a migraine headache. Figure 1.2 illustrates the closed discovery setting on Swanson's task of finding the terms linking the *migraine* and *magnesium* domains. It is important to note that Swanson's literature mining results have been later confirmed by laboratory and clinical investigations. This well-known example has become the gold standard

[5] In the ABC model, uppercase letter symbols A, B and C are used to represent domains (sets of terms), and lowercase symbols a, b and c to represent concepts (individual terms).

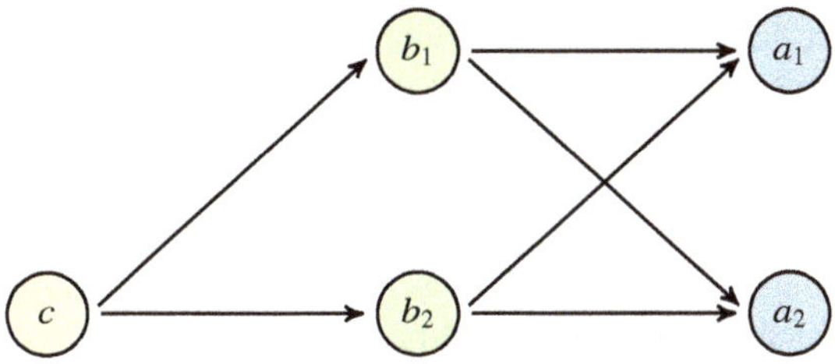

Fig. 1.4 In open discovery, only the starting concept c is given and the goal is to identify target concepts $a_1, a_2, \ldots, a_m$ through intermediate concepts $b_1, b_2, \ldots, b_n$.

in the literature mining field and has been used as a benchmark in several studies (Juršič et al., 2012a; Lindsay & Gordon, 1999; Srinivasan, 2004; Weeber et al., 2001).

Open discovery. The open discovery setting has been also widely explored (Weeber et al., 2001). The open discovery setting is illustrated in Figure 1.4. Open discovery is in essence the approach employed for generating new hypotheses. Let us describe it in terms of Swanson's (1986) ABC model. Let the starting point of research be term c, let C be the literature about c, and let us assume that domain A consists of the literature that might contain some—yet unknown—connections to C. The goal is to find A. This is done

> via some intermediate literature (B) toward an unknown destination A... Success depends entirely on the knowledge and ingenuity of the searcher... (Swanson, 1990)

In effect, in open discovery one is interested in concept c, and tries to find some target term a in A, to formulate a possible hypothesis. For example, in the case of diseases or disorders, such a hypothesis could be a suggestion of treatment for a disease c by drug a, which is known to be effective in treating diseases in domain A, but which could be repurposed for treating disease c in C, possibly explained through some intermediate phenomenon b in B.
Note that an early approach proposed by Lindsay and Gordon (1999), using lexical statistics to determine relative frequencies of words and phrases in the search for words on the top of the ranked list of terms, failed when applied to Swanson's first discoveries and extensive analysis had to be based on human knowledge and judgment.

Inspired by the early work of Swanson, literature mining approaches were further developed and successfully applied to different problems, such as finding associations between diseases and chemicals (Yetisgen-Yildiz & Pratt, 2006), drug repurposing (Weeber et al., 2003), and many others. The open discovery approach developed by Srinivasan (2004) relied on Medical Subject Headings (MeSH). Yetisgen-Yildiz and Pratt (2006) proposed a LBD system called LitLinker, which mines biomedical literature with knowledge-based and statistical methods. All these systems use MeSH descriptors (Nelson et al., 2001) as a representation of scientific medical documents, instead of using title, abstract, or full-text words. However, a problem is that MeSH indexers normally use only the most specific vocabulary to describe the topics discussed in the given documents and therefore some significant terminology from the content of documents may not be covered. Holzinger et al. (2013) described several web-based tools for the analysis of biomedical literature, which include the analysis of terms (biomedical entities such as disease, drugs, genes, proteins and organs) and provide concepts associated with a given term. The work of Kastrin et al. (2014) is complementary to other LBD approaches, as it uses different similarity measures (such as common neighbors, Jaccard index, and preferential attachment) for link prediction of implicit relationships in the Semantic MEDLINE network.

Comprehensive surveys of LBD approaches in the biomedical domain can be found in the works of several authors:

- Ahmed (2016) provided the first attempt to systematically classify LBD methods and approaches through the lens of information retrieval. The paper identified three approaches that were most often used as a basis for LBD: (i) vector space model, (ii) probabilistic methods, and (iii) inference network.
- Sebastian et al. (2017) published the first extensive review paper on LBD. The authors provided in-depth discussion on a broad palette of existing LBD approaches and offered performance evaluations on some recent emerging LBD methodologies, providing a general unifying framework for LBD. Traditional approaches currently dominate LBD research. They comprise methods that rely on (i) lexical statistics, (ii) knowledge-based techniques, and (iii) visualization techniques.
- Nearly at the same time, Henry and McInnes (2017) published a high-level overview with an extensive survey of LBD methods including (i) co-occurrence models, (ii) semantic models, and (iii) distributional semantics models.
- Gopalakrishnan et al. (2019) presented an attempt to consolidate the evolution of discovery approaches and computational techniques in the area of LBD. The authors discussed the main LBD approaches by critically analyzing selected influential works and also provided a broad classification schema including (i) co-occurrence-based approaches, (ii) semantic relation-based approaches, (iii) graph-based approaches, and (iv) hybrid approaches.
- Thilakaratne et al. (2019a, 2019b) analyzed the methodologies used in LBD via a classification scheme, as well as the LBD research trends through the analysis of LBD publications, complemented by the manual analysis of a large number of published LBD articles.
- Kastrin and Hristovski (2021) presented the scientometric analysis of LBD research papers to structure LBD knowledge and analyze the evolution of the LBD literature.

LBD has gained considerable attention primarily in the biomedical domain. This dominance can be attributed to two fundamental characteristics of biomedical research that make LBD particularly effective and widely accepted within this research field.

First, biomedicine benefits from the availability of open-access data and well-curated databases. Life sciences generate a large number of scientific articles that are freely available through the PubMed bibliographic service. In addition, a plethora of other auxiliary natural language processing (NLP) resources including controlled vocabularies such as MeSH and ontologies such as Unified Medical Language System (UMLS) provide a standardized framework for text mining tasks (e.g., entity recognition, information extraction).

Second, biomedical research is inherently interdisciplinary and spans multiple disciplines (e.g., genetics and pharmacology), leading to vast but fragmented knowledge. This offers a natural playground for LBD, helping to bridge the gaps between disparate parts of knowledge by finding connections that might not be obvious within individual subfields.

1.3 Underlying principles and theories

This section outlines some of the principles and theories underlying modern literature-based discovery approaches, including the design principles used in creativity support tools, conceptual representations in concept creation and discovery, transformative discovery, and bisociative reasoning.

1.3.1 Design principles of creativity support tools

LBD typically not an automated process and assumes human-in-the-loop. The process relies on using online LBD tools, using Human-Computer Interaction (HCI) interfaces. From the HCI perspective, LBD tools can be broadly considered as creativity support tools (CSTs), whose design aims to develop improved software and user interfaces that empower users to be both more productive and more innovative (Swanson, 1990).

A structured set of design principles for CSTs by Shneiderman (2007, 2009) is outlined below.

Support exploration. To be successful in discovery and innovation, users should have access to improved search services that provide numerous tools to organize search results by ranking, clustering, and partitioning with ample tools for annotation, tagging, and marking.

Enable collaboration. Although the actual discovery moments in innovation can be very personal, the processes that lead to them are often highly collaborative.

Provide rich history-keeping. The benefits of rich history keeping are that users have a record of which alternatives they have tried, they can compare the many alternatives, and they can go back to earlier alternatives to make modifications.

Design with low thresholds, high ceilings, and wide walls. CST should have a steep learning curve for novices (low threshold), yet provide sophisticated functionalities that experts need (high ceilings), and also deliver a wide range of supplementary services to choose from (wide walls).

These principles were largely followed in the LBD tools described in this monograph, including the CSTs OntoGen (Fortuna et al., 2006), CrossBee (Juršič et al., 2012a; Juršič et al., 2012b), and RaJoLink (Petrič et al., 2009).

1.3.2 Concept creation and discovery

LBD can be broadly considered as part of computational concept creation and discovery research. The typology of this research, proposed by Xiao et al. (2019), may help us better understand the positioning of LBD in this research context. In our view, LBD can be seen as a form of concept creation and discovery, falling into the following categories described below: concept extraction, concept induction and concept space exploration.

A well-accepted categorization of the different types of creativity was provided by Boden (1990), distinguishing between combinatorial reuse of existing ideas, exploratory search for new ideas, and transformational creativity, where the search space is also subject to change.

Although Boden (1990) does not give a computational account of how concepts or ideas could be created, her taxonomy was used as a basis for characterizing various possible approaches to concept creation and discovery by Xiao et al. (2019), who extended this categorization with approaches based on extraction and induction, basing their taxonomy of concept creation approaches on the type of input they take.

- *Concept extraction* transforms an existing representation of a concept to a different representation,
- *Concept induction* generalizes a set of instances of concepts to more general concepts,
- *Concept recycling* modifies existing concepts to new ones,
- *Concept space exploration* takes a search space of concepts as input, emphasizing the concept traversal strategy rather than modifying the input.

Selected concept creation approaches are explained below.

Concept extraction is the task of extracting and transforming a conceptual representation from an existing representation of a concept to a different representation of the same idea, where the resulting extracted representation is more concise, explicit, and centralized. Concept extraction is frequently applied to textual corpora, such as extracting semantic relations between concepts. For an overview of information extraction methods, see e.g., the work of Sarawagi (2008).

Concept induction is based on instances of concepts from which the concepts are learned. Drawing from the field of machine learning and data mining, two major types of inductive concept creation can be identified:

Concept learning is a supervised activity, where a description of a concept is formed inductively from (positive and negative) examples of the concept, see e.g., the work of Kotsiantis (2007).

Concept discovery is an unsupervised inductive activity where the memberships of instances in concepts are not known in advance, but methods such as clustering are used to generalize natural concepts from given examples, see e.g., the work of Jain et al. (1999).

Concept recycling is the creative reuse of existing concepts (Aamodt & Plaza, 1994), which includes two typical methods:

Concept mutation modifies a given concept by adding, removing, or changing something in it, where two common mutational operations are *generalization* and *specialization* of existing concepts (Osborn, 1953).

Concept combination where two or more existing concepts are combined to form a new one, such as in *conceptual blending*, where 'computer program' and 'virus' can form a conceptual blend of 'computer virus'.

Concept space exploration takes as input a search space of possible new concepts and locates interesting concepts in the search space. The space can be specified either declaratively or procedurally. An opportunity for exploratory creativity is to discover novel concepts in word association graphs by exploiting graph structure measures.

The above typology is based on the differences of the approaches in terms of their input representations and can thus help characterize different settings where concept creation and discovery are used. A creative system can also adjust its own operation, e.g., by modifying its search space, its objectives, or the search methods (Boden, 1990; Wiggins, 2006).

1.3.3 Transformative discovery

The central premise of many theories and models of scientific discovery is to describe the evolution of research ideas, explain the process of creativity, and elucidate scientific revolutions. One of the more promising theories in this area, the Theory of Transformative Discoveries (TTD), proposed by Chen et al. (2009), provides an innovative conceptual framework to explain new relationships between disparate fields of knowledge. TTD draws heavily on the model of structural holes, proposed by Burt (1995), which refers to the gap in the network structure.

In Burt's model, a structural hole can control the flow of information between the two, originally unconnected parts of the network. For example, if node b has direct connections with a group of nodes A and C, where the nodes in A and C are not directly connected, node b can be considered a bridge between nodes A and C. This brokerage mechanism, which is considered a generic discovery mechanism in the TTD, is the generic discovery mechanism also used in LBD.

The second part of TTD refers to Information Foraging and explains the process of knowledge diffusion. Regarding Information Foraging, Chen et al. (2009) introduced two measures, namely betweenness centrality (structural measure) and citation burstness (temporal measure). TTD uses a combination of both measures to define which research paper receives a high impact.

1.3.4 Bisociative reasoning

In computational creativity, associations play a prominent role. Merriam-Webster dictionary entry[6] defines *association* as follows:

> something linked in memory or imagination with a thing or person; the process of forming mental connections or bonds between sensations, ideas, or memories.

The notion of association assumes a connection between two concepts or symbols c_1 and c_2, but does not assume any specific conditions on c_1 and c_2, i.e. the nature of the connection between two associated concepts is not specified, as in the case of e.g., semantic relations, semantic networks, taxonomies, or ontologies.

Koestler (1964) investigated cross-domain associations as an important element of creativity, calling them *bisociations*. Bisociations, i.e. associations spanning over different contexts (domains/categories/classes), which may be very useful in discovering or creating analogies and metaphors, are particularly interesting for literature-based discovery. Bisociation is in essence the result of a creative endeavor. In order to connect pieces of information from previously unrelated domains, a person must activate some form of creative mechanism. This creative aspect is what allows one to go beyond one-dimensional associations. This has been recognized in several psychocognitive theories related to creativity, which share the principle that a strong connection exists between creative activity and the ability to establish relations between seemingly unrelated domains.

Divergence can nevertheless be achieved to a certain degree by cross-domain reasoning in a multi-domain database. Such a model must provide mechanisms for mapping concepts and transferring meanings. According to Koestler (1964), in addition to *metaphor*, the well-known examples of mechanisms that can be used in cross-domain knowledge transfer are *analogy* and *bisociation*.

1.3.5 Types of bisociations

While an association bridges concepts within a given domain, according to Berthold (2012), a bisociation bridges two unconnected (or only very sparsely connected) domains. Two concepts c_1 and c_2, from two different contexts D_1 and D_2 are bisociated, if there is no direct, obvious evidence linking them, one has to cross contexts to find the link, and this new link provides novel insights into the domains. As an example, take a simple bisociation between natural *evolution* and *computing*, used when coining the term *evolutionary computing*. Although the principle of bisociation cannot be equated with creativity in general, it is a special case of *combinatorial creativity*, which refers to novel combinations of ideas.

For convenience, to comply with Swanson's ABC model, let us replace c_1 by a, and c_2 by c in the schematic presentation of bisociation in Figure 1.5. Furthermore, a

[6] http://www.merriam-webster.com/dictionary/association

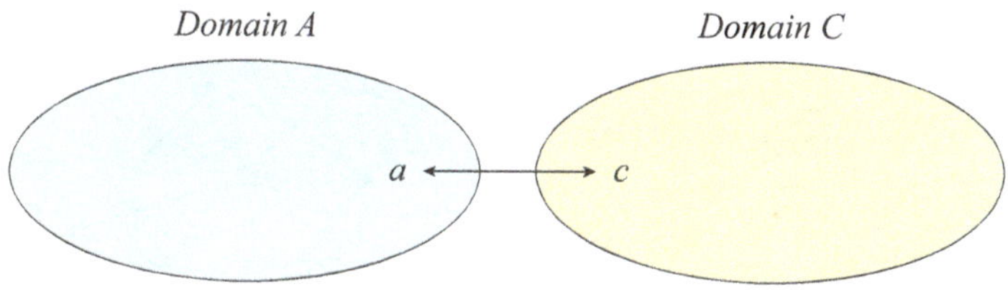

Fig. 1.5 Bisociation between concept a in domain A and concept c in domain C.

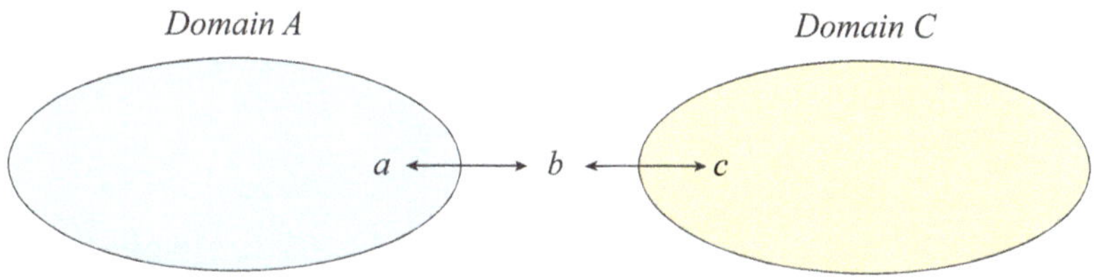

Fig. 1.6 Bisociation between concepts a and c using a bridging concept b.

bisociative connection between a and c may be represented together with a *bridging* concept b, which has links to both a and c (see Figure 1.6).

Starting from the Koestler's concept of bisociation (Koestler, 1964), concrete bisociative patterns that are searched for in bisociative knowledge discovery include: bridging concepts, bridging graphs, and bridging by structural similarity (Kötter & Berthold, 2012):

Bridging concepts. This is the most natural type of bisociation: a concept connecting two domains. In practice, different literatures from different domains are explored, and some terms connecting the two are found. This is the kind of pattern originally explored by Swanson. These connecting terms allow us to corroborate hypotheses linking the two domains. A bridging concept at the intersection of two domains A and C is illustrated in Figure 1.6 and Figure 1.7.

Bridging graphs. More complex bisociations are modeled by bridging graphs, in a network representation. This is similar to bridging concepts, but in this case, what connects two different domains is a subset of *related* concepts, illustrated in Figure 1.8.

Bridging structural similarity. This is the most complex kind of bisociation, where again in a network representation, subsets of concepts in each domain share structural similarities, illustrated in Figure 1.9.

Bisociations based on structural similarity are represented by relations and/or subgraphs of two different, structurally similar domains (Kötter & Berthold, 2012), illustrated in Figure 1.8. This type of bisociation is, according to Kötter and Berthold (2012), the most abstract pattern with the potential for new cross-domain discoveries that, for example, vertex similarity methods can identify.

Note that the scientific question addressed in Chapter 9 is whether embeddings can be used as means to discover *relational bisociations* (Lavrač et al., 2020), a special case of bridging by structural similarity. This type of bisociation is illustrated in Figure 1.9.

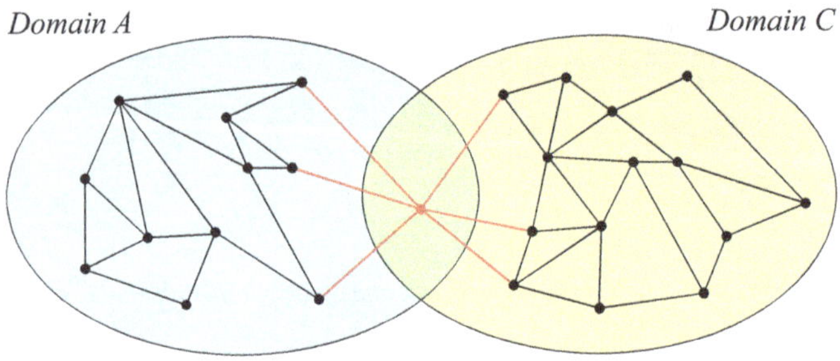

Fig. 1.7 Bridging concept in the intersection of two literature domains A and C.

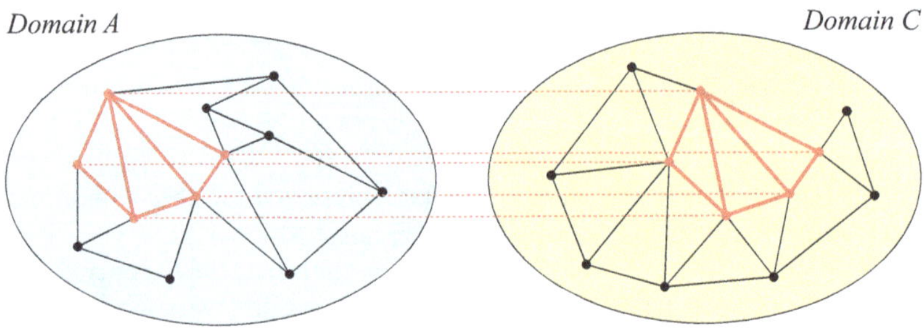

Fig. 1.8 Bridging by graph similarity (Kötter & Berthold, 2012).

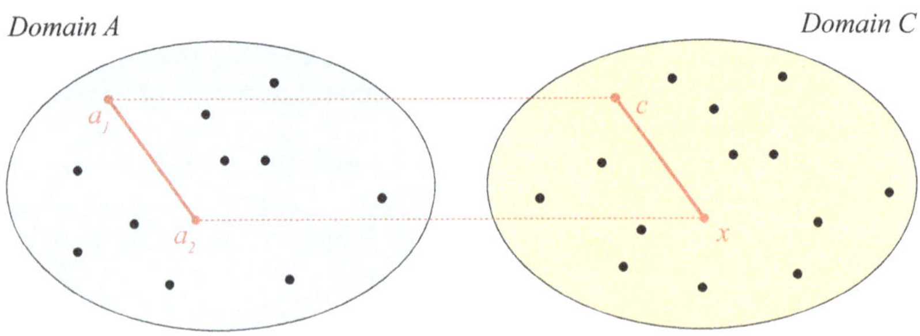

Fig. 1.9 Concept of bridging by relational bisociation (Lavrač et al., 2020).

1.3.6 Relating Swanson's ABC model and bisociative LBD

Bisociative knowledge discovery (Berthold, 2012) can be considered a data mining task in which two or more domains of interest are searched for bisociative links or bridging concepts (i.e. single cross-context terms). Note that a single domain in this context does not necessarily refer to a single feature space. Instead, we use this term to indicate that the analyzed objects represent all features of one—more or less specific—aspect, even if there are multiple representations of the same object space (multiview learning, parallel universes, and redescription mining are the well-known techniques that deal with multiple representations of objects in the same discourse domain). In contrast, bisociative knowledge discovery searches for connections between different discourse domains using either the same representation of different domains or different domain representations, where—according to Berthold (2012)—bridging concepts can be detected as nodes bridging different graphs, as subgraphs connecting different graphs, as bridging connections in the sense of graph similarity, or as bridging concepts occurring in different document corpora. The latter is referred to as *bridging term discovery*.

Let us summarize the relationship between bisociative knowledge discovery and Swanson's ABC model for LBD, previously published by Petrič et al. (2012) and Lavrač et al. (2019), where the particular focus of interest is the relationship between Koestler's bisociative link discovery framework and Weeber's closed discovery framework, summarized in Table 1.1.

Table 1.1 Unifying Koestler's and Swanson's models of creative knowledge discovery (Petrič et al., 2012).

Koestler's model	Swanson's and Weeber's model
bisociative link discovery process	closed discovery process
frames of reference (contexts) M_1 and M_2	domains of interest A and C
$l \in M_1 \cap M_2$	bridging term b $\in$ terms(A) $\cap$ terms(C)

Similarly to bisociations, which according to Koestler are the result of processes of mind when making new associations between concepts from usually separated contexts, LBDs in Swanson's ABC model are a result of uncovering links between concepts a and c from disjoint literatures A and C. In terms of Koestler's model, the two domains A and C, investigated in the closed LBD framework, correspond to the two habitually incompatible frames of reference, M_1 and M_2. Moreover, the bridging terms $b_1, b_2, \ldots, b_n$ that are common to literature A and C, clearly correspond to Koestler's notion of a situation or idea l, which is not merely linked to one associative context but bisociated with two contexts M_1 and M_2.

1.4 Evaluation of LBD approaches

Evaluating LBD systems presents significant challenges due to the lack of consensus among researchers on the most effective assessment methods. Two key unresolved issues in the LBD literature further complicate this evaluation:

1. determining whether novel discoveries are the result of infrequent or commonly co-occurring terms, and
2. automating the evaluation processes effectively.

The subsequent sections outline two evaluation strategies that are widely adopted within the LBD community and particularly relevant to the discussions in later chapters of this book: the *discovery replication* strategy and the *time-slicing approach.*

1.4.1 Discovery replication

Discovery replication aims to recreate historical scientific discoveries. A common method in LBD involves replicating known associations, such as Swanson's links between magnesium deficiency and migraines or Raynaud's disease and fish oil (refer to the benchmark datasets in Section 4.1). This strategy uses only the knowledge available prior to the publication of the original discovery. A system is deemed successful if it can identify and appropriately rank the target discovery in its output.

This approach is computationally straightforward and can be automated across various LBD systems, providing a clear demonstration of a system's replication capabilities. However, it typically focuses on a limited number of discoveries, which might not represent the system's overall performance. Even small modifications in the LBD process, such as altering the literature retrieval strategy, can significantly affect the outcomes, introducing a high risk of overfitting and making the ranking metric unstable. Thus, while discovery replication offers insights into the efficacy of the entire LBD process, it does not address the performance of individual components.

However, discovery replication focuses only on a single discovery. Even when multiple historical discoveries have been replicated, their scope remains limited. Only a small adjustment in the LBD process (e.g., the literature retrieval strategy to extract terms that link concepts c and b) can cause significant variations in ranking. The risk of overfitting is considerably high, making the rank metric unstable. Discovery replication assesses the quality of an LBD system in a discrete manner, by evaluating the performance of the entire LBD process rather than its individual components.

1.4.2 Time-based evaluation

Time-based evaluation involves partitioning the dataset into training and test sets based on a predetermined cutoff date. The training set, which precedes the cutoff, is used to generate hypotheses. In contrast, the test set, which follows the cutoff, helps establish a gold standard for future discoveries. This gold standard is formulated using the selected LBD approaches, such as analyzing term co-occurrences or extracting semantic relationships that appear in the post-cutoff dataset but not in the pre-cutoff dataset, indicating the emergence of new knowledge. The evaluation then employs standard metrics like precision, recall, and F-measure to assess the accuracy of both predicted and validated discoveries.

1.5 Outline of the monograph and the approach to code reuse

This monograph serves as an introduction to LBD, focusing on the LBD approaches and implementations co-developed by the authors of the monograph. The monograph text is structured in terms of LBD field categorization, which has been largely inspired by the approach to field contextualization in terms of cognitive representation levels proposed by Gärdenfors (2000), and the taxonomy of concept creation approaches based on the typology of input data proposed by Xiao et al. (2019) (see Section 1.3.2).

As most human knowledge analyzed in LBD is stored in textual format, while the most powerful analysis tools take as input spatial vector representations, this explains a plethora of techniques that transform human knowledge in textual format to the spatial representation space, including the so-called Bag of Words representations regularly used in text mining-based approaches (see Section 3.3). In addition to human knowledge in the form of text documents, available in public literature resources (see Section 2.4.1), some of the approaches take into account also explicitly encoded semantic knowledge in the form of taxonomies and ontologies (see Section 2.4.2). Finally, numerous LBD approaches are based on data representation with network or graph representation models (see Section 3.4.1). More recently, compact spatial vector representations of texts became most popular, which are exploited in text embeddings (see Section 3.5) and LBD approaches based on large language models (LLMs) (see Section 3.6). We used this representation framework also as a guideline for structuring the presentation of the approaches covered by this monograph, in accordance with the following top-level categorization of LBD approaches: text mining-based, concept-based, network-based, embeddings-based and LLM-based LBD approaches.

The monograph is structured as follows. After a brief survey of the LBD field in Chapter 2, an overview of related background technologies in Chapter 3, and the presentation of benchmark data sets and the approach to experiment replicability in Chapter 4. The main body of the monograph covers the following LBD approaches:

- Selected *text mining-based* LBD approaches using standard Bag of Words document representations are presented Chapter 5, Chapter 6 and Chapter 7 (Section 7.3).
- A selected *concept-based* LBD approach is presented in Chapter 7 (Section 7.2).
- A selected *network-based* LBD approach is presented in Chapter 8.
- A selected *embeddings-based* approach is presented in Chapter 9.
- An illustrative *LLM-based* approach to LBD domain conceptualization is presented in Chapter 10, where the latter includes reflections upon the LBD research avenues opened by the recent LLMs and outlines some of the lessons learned from our own past LBD research.

In writing this monograph, we pursued an open science approach, briefly outlined in Chapter 4. To illustrate the use of LBD techniques described in the monograph, many of the described LBD approaches have been implemented as illustrative LBD pipelines. It was not our intention to fully reimplement the original approaches to LBD, but rather to illustrate some of their crucial ingredients as components of the implemented pipelines.

The pursued open-science approach is beneficial due to its potential for code reuse and experiment replicability. The backbone of the monograph's experimental infrastructure is two-fold:

- The available benchmark datasets, which are presented in Section 4.1.
- The implemented Jupyter Python notebook solutions, illustrating selected concepts and LBD pipelines presented in this monograph, with details of the installation of developed Jupyter Python notebooks, are outlined in Section 4.2.

Except for Chapters 1–4, most of the other chapters conclude with a dedicated section *Tutorials in Python*, where Jupyter Python notebooks are used to demonstrate selected LBD pipelines described in a given chapter, allowing the reader to run the described methods on illustrative examples. The notebooks provide a set of examples that have been carefully chosen to be easy to understand but are not trivial and can be used as recipes or starting points in developing more complex LBD scenarios.

References

Aamodt, A., & Plaza, E. (1994). Case-based reasoning: Foundational issues, methodological variations, and system approaches. *AI Communications*, *7*(1), 39–59.

Ahmed, A. (2016). Literature-based discovery: Critical analysis and future directions. *International Journal of Computer Science and Network Security*, *16*(7), 11–26.

Berthold, M. R. (Ed.). (2012). *Bisociative knowledge discovery*. Springer.

Boden, M. A. (1990). *The creative mind: Myths and mechanisms*. Basic Books.

Bruza, P., & Weeber, M. (Eds.). (2008). *Literature-based discovery*. Springer.

Burt, R. S. (1995). *Structural holes: The social structure of competition*. Cambridge: Harvard University Press.

Chen, C., Chen, Y., Horowitz, M., Hou, H., Liu, Z., & Pellegrino, D. (2009). Towards an explanatory and computational theory of scientific discovery. *Journal of Informetrics*, *3*(3), 191–209.

Cheng, L., Lin, H., Zhou, F., Yang, Z., & Wang, J. (2014). Enhancing the accuracy of knowledge discovery: A supervised learning method. *BMC Bioinformatics*, *15*(Suppl 12), S9.

Dubitzky, W., Kötter, T., Schmidt, O., & Berthold, M. (2012). Towards creative information exploration based on Koestler's concept of bisociation. In M. R. Berthold (Ed.), *Bisociative knowledge discovery* (pp. 11–32). Springer.

Fortuna, B., Grobelnik, M., & Mladenić, D. (2006). Semi-automatic data-driven ontology construction system. *Proceedings of the 9th International Multiconference Information Society*, 223–226.

Gärdenfors, P. (2000). *Conceptual spaces: The geometry of thought*. The MIT Press.

Gopalakrishnan, V., Jha, K., Jin, W., & Zhang, A. (2019). A survey on literature based discovery approaches in biomedical domain. *Journal of Biomedical Informatics*, *93*, 103141.

Govaert, G., & Nadif, M. (2014). *Co-clustering: Models, algorithms and applications*. Wiley.

Henry, S., & McInnes, B. T. (2017). Literature based discovery: Models, methods, and trends. *Journal of Biomedical Informatics*, *74*, 20–32.

Holzinger, A., Yildirim, P., Geier, M., & Simonic, K.-M. (2013). Quality-based knowledge discovery from medical text on the web. In *Quality Issues in the Management of Web Information* (pp. 11–13). Springer.

Jain, A. K., Murty, M. N., & Flynn, P. J. (1999). Data clustering: A review. *ACM Computing Surveys*, *31*(3), 264–323.

Juršič, M., Cestnik, B., Urbančič, T., & Lavrač, N. (2012a). Cross-domain literature mining: Finding bridging concepts with CrossBee. *Proceedings of the 3rd International Conference on Computational Creativity*, 33–40.

Juršič, M., Cestnik, B., Urbančič, T., & Lavrač, N. (2012b). Bisociative literature mining by ensemble heuristics. In M. R. Berthold (Ed.), *Bisociative knowledge discovery* (pp. 338–358). Springer.

Kastrin, A., & Hristovski, D. (2021). Scientometric analysis and knowledge mapping of literature-based discovery (1986–2020). *Scientometrics*, *126*(2), 1415–1451.

Kastrin, A., Rindflesch, T. C., & Hristovski, D. (2014). Link prediction on the semantic MEDLINE network. *Proceedings of the 17th International Conference on Discovery Science*, 135–143.

Koestler, A. (1964). *The act of creation*. Hutchinson.

Kotsiantis, S. B. (2007). Supervised machine learning: A review of classification techniques. *Proceedings of the 2007 Conference on Emerging Artificial Intelligence Applications in Computer Engineering*, 3–24.

Kötter, T., & Berthold, M. (2012). From information networks to bisociative information networks. In M. R. Berthold (Ed.), *Bisociative Knowledge Discovery* (pp. 33–50). Springer.

Lavrač, N., Juršič, M., Sluban, B., Perovšek, M., Pollak, S., Urbančič, T., & Cestnik, B. (2019). Bisociative knowledge discovery for cross-domain literature mining. In T. Veale & F. A. Cardoso (Eds.), *Computational Creativity: The Philosophy and Engineering of Autonomously Creative Systems* (pp. 121–139). Springer.

Lavrač, N., Martinc, M., Pollak, S., Pompe Novak, M., & Cestnik, B. (2020). Bisociative literature-based discovery: Lessons learned and new word embedding approach. *New Generation Computing*, *38*(3), 773–800.

Lindsay, R. K., & Gordon, M. D. (1999). Literature-based discovery by lexical statistics. *Journal of the American Society for Information Science and Technology*, *50*(1), 574–587.

Nelson, S. J., Johnston, W. D., & Humphreys, B. L. (2001). Relationships in Medical Subject Headings (MeSH). In C. A. Bean & R. Green (Eds.), *Relationships in the organization of knowledge* (pp. 171–184). Springer.

Osborn, A. F. (1953). *Applied imagination*. Scribner.

Petrič, I., Cestnik, B., Lavrač, N., & Urbančič, T. (2012). Outlier detection in cross-context link discovery for creative literature mining. *The Computer Journal*, *55*(1), 47–61.

Petrič, I., Urbančič, T., Cestnik, B., & Macedoni-Lukšič, M. (2009). Literature mining method RaJoLink for uncovering relations between biomedical concepts. *Journal of Biomedical Informatics*, *42*(2), 219–227.

Sarawagi, S. (2008). Information extraction. *Foundations and Trends in Databases*, *1*(3), 261–377.

Sebastian, Y., Siew, E.-G., & Orimaye, S. O. (2017). Emerging approaches in literature-based discovery: Techniques and performance review. *The Knowledge Engineering Review*, *32*, e12.

Shneiderman, B. (2007). Creativity support tools: Accelerating discovery and innovation. *Communications of the ACM*, *50*(12), 20–32.

Shneiderman, B. (2009). Creativity support tools: A grand challenge for HCI researchers. In *Engineering the User Interface* (pp. 1–9). Springer.

Smalheiser, N. R., & Swanson, D. R. (1998). Using ARROWSMITH: A computer-assisted approach to formulating and assessing scientific hypotheses. *Computer Methods and Programs in Biomedicine*, *57*(3), 149–154.

Srinivasan, P. (2004). Text mining: Generating hypotheses from MEDLINE. *Journal of the American Society for Information Science and Technology*, *55*(5), 396–413.

Swanson, D. R. (1986). Undiscovered public knowledge. *The Library Quarterly*, *56*(2), 103–118.

Swanson, D. R. (1988). Migraine and magnesium: Eleven neglected connections. *Perspectives in Biology and Medicine*, *78*(1), 526–557.

Swanson, D. R. (1990). Medical literature as a potential source of new knowledge. *Bulletin of the Medical Library Association*, *78*(1), 29–37.

Swanson, D. R., Smalheiser, N. R., & Torvik, V. I. (2006). Ranking indirect connections in literature-based discovery: The role of Medical Subject Headings (MeSH). *Journal of the American Society for Information Science and Technology*, *57*(11), 1427–1439.

Thilakaratne, M., Falkner, K., & Atapattu, T. (2019a). A systematic review on literature-based discovery: General overview, methodology, & statistical analysis. *ACM Computing Surveys*, *52*(6), 129:1–129:34.

Thilakaratne, M., Falkner, K., & Atapattu, T. (2019b). A systematic review on literature-based discovery workflow. *PeerJ Computer Science*, *5*, e235.

Weeber, M., Klein, H., de Jong-van den Berg, L., Vos, R., et al. (2001). Using concepts in literature-based discovery: Simulating Swanson's Raynaud–fish oil and migraine–magnesium discoveries. *Journal of the American Society for Information Science and Technology*, *52*(7), 548–557.

Weeber, M., Vos, R., Klein, H., de Jong-van den Berg, L. T., Aronson, A. R., & Molema, G. (2003). Generating hypotheses by discovering implicit associations in the literature: A case report of a search for new potential therapeutic uses for thalidomide. *Journal of the American Medical Informatics Association*, *10*(3), 252–259.

Wiggins, G. A. (2006). A preliminary framework for description, analysis and comparison of creative systems. *Journal of Knowledge Based Systems*, *19*(7), 449–458.

Xiao, P., Toivonen, H., Gross, O., Cardoso, A., Correia, J., Machado, P., Martins, P., Oliveira, H. G., Sharma, R., Pinto, A. M., Díaz, A., Francisco, V., Gervás, P., Hervás, R., León, C., Forth, J., Purver, M., Wiggins, G. A., Miljković, D., . . .

Battersby, S. (2019). Conceptual representations for computational concept creation. *ACM Computing Surveys*, *52*(1), 1–33.

Yetisgen-Yildiz, M., & Pratt, W. (2006). Using statistical and knowledge-based approaches for literature-based discovery. *Journal of Biomedical Informatics*, *39*(6), 600–611.

Chapter 2
History, Resources and Tools

This chapter provides an overview of the selected approaches to LBD. It begins with a historical survey of the field in Section 2.1, and continues with a summary of LBD review articles in Section 2.2. Section 2.3 presents the scientometric and longitudinal keyword and topic analysis of the LBD field. The chapter then presents typical LBD resources in Section 2.4, followed by the presentation of selected LBD tools in Section 2.5.

2.1 Historical sketch

Although the link between fish oil and Raynaud's disease was discovered by Swanson (1986a) by chance during a manual search of the literature, researchers later built on the assumption that LBD is generally a rational and explainable process, to a large extent driven by serendipity.

Traditionally, *discovery by serendipity* is described as an accidental but successful finding of a new piece of knowledge while searching for something else. Famous examples of serendipitous discoveries in modern science include Newton's explanation of universal gravitation and Flemming's discovery of penicillin. Recent empirical findings suggest that serendipity plays a key role in scientists' work (García, 2009). According to Fine and Deegan (1996), there are three general determinants of serendipity:

1. temporal happenstance,
2. analytical or active learning, and
3. relational network.

The next important concept is *complementarity*, which refers to the relationship between two objects. Although Swanson used the term complementarity in his initial definition of the LBD field (Swanson, 1986b), he later understood complementarity as the human ability to recognize the relationships between scientific ideas and in this context also redefined his initial definition of the LBD process as follows (Swanson, 1986b):

> The goal of LBD, in my opinion, should be to support and enhance human ability by focusing on the key problems of finding promising pairs of scientific articles that can serve as a stimulus

N. Lavrač et al., *Bisociative Literature-Based Discovery*,
https://doi.org/10.1007/978-3-031-96863-1_2

and on identifying associated literature structures. In any event, it is plausible to assume that two articles randomly selected from a vast literature would have almost no chance of being complementary, so we need a search process that combines human knowledge and judgment with computer speed and data capacities.

Since the mid-1990s, various approaches, techniques, and tools have been developed for LBD. According to Gopalakrishnan et al. (2019), the main research lines in LBD can be split into the following broad categories: (i) co-occurrence-based, (ii) semantic relations-based, (iii) graph-based, and (iv) hybrid approaches. In addition, we added another category, (v) embeddings-based and Large Language Models (LLM)-based approaches, to cover newer approaches covered in this monograph. Following this categorization, we outline the well-known LBD techniques without addressing their specific characteristics, their applications, or the intricacies of their implementation.

Co-occurrence-based approaches. While Swanson (1986a) made his breakthrough discovery by hand, using a trial-and-error approach, computational LBD research began with the notion of co-occurrence, which the authors used to imitate interactions between knowledge entities, mirroring the process of knowledge discovery in humans. Formally, co-occurrence was defined as the pattern of association between two items occurring in the same document, paragraph, or sentence. In representing explicit knowledge through co-occurrence, it is argued that the presence or absence of a link reflects the actual relatedness or unrelatedness between the two phenomena in nature. An initial attempt to automate the LBD process was made by Swanson and Smalheiser (1997) who built an interactive program called Arrowsmith. Each MEDLINE document was represented by its title, and weighted co-occurrences between single words (and *n*-grams in subsequent versions) denoted their co-occurrence. The authors proposed many heuristics (e.g., a stop list of uninformative phrases) to reduce human involvement in the discovery process. Co-occurrence has been employed in numerous LBD applications. For example, Pyysalo et al. (2019) developed a web application called LION LBD and integrated a range of co-occurrence metrics to analyze implicit links between knowledge entities. However, the reader should keep in mind that co-occurrence alone cannot capture the full semantic meaning of the given knowledge entity and is hence not always meaningful. In this monograph, a co-occurence-based text mining LBD approach is presented Chapter 5.

Semantic relations-based approaches. Compared to Swanson's work, a step further was taken by Weeber et al. (2001), who expanded the document representation framework to the Title and Abstract fields and, more importantly, normalized the plain text to concepts defined in Unified Medical Language System (UMLS)[1] Metathesaurus, using the MetaMap tool. The authors demonstrated the efficacy of UMLS normalization, which usually reduces the dimensionality of the input space by several orders of magnitude. In addition to employing the entire concept space of the UMLS Metathesaurus, the relevant normalizations can also be done with the much smaller Medical Subject Headings (MeSH)[2] vocabulary. The co-occurrence-based approach proposed by Weeber et al. (2001), discussed above, introduced semantic information to some extent, filtering out uninteresting bridging concepts based on the semantic type of the

[1] https://lhncbc.nlm.nih.gov/semanticnetwork/

[2] nlm.nih.gov/mesh

concept (e.g., Disease or Syndrome, Pharmacologic Substance). This simple technique, called *semantic filtering*, has been commonly used to reduce the complexity of the hypothesis space and therefore improve the accuracy and interpretability of the discovery process. A computational alternative to semantic filtering was developed by Srinivasan (2004), who introduced the technique of topic profiling. However, the era of semantic LBD approaches began with the use of SemRep (Rindflesch & Fiszman, 2003), a standard NLP tool (briefly described in Section 2.5.1) that extracts semantic relations from PubMed titles and abstracts, intensified with the release of the SemMedDB repository (Kilicoglu et al., 2020).

Several researchers have continued Swanson's and Weeber's line of research. The open discovery approach developed by Srinivasan and Libbus (2004) is heavily based on MeSH. Yetisgen-Yildiz and Pratt (2006) proposed a biomedical literature-based discovery system, called LitLinker, utilizing knowledge-based and statistical methods. An online literature-based discovery tool called BITOLA, designed by Hristovski et al. (2005), uses association rule mining techniques to find implicit relations between biomedical terms. Weeber (2007) developed Literaby, a concept-based Natural Language Processing (NLP) tool. The units of analysis essential for the approach are UMLS Metathesaurus concepts. All above systems employ MeSH descriptors (Nelson et al., 2001) as a representation of scientific medical articles, however, the problem arises since MeSH indexers typically use only the most specific vocabulary to describe the topic discussed in a document and therefore some significant terminology from the document content may not be covered. In this monograph, a selected concept-based LBD approach is presented in Section 7.2 of Chapter 7.

Network-based approaches. Network-based methods represent the third high-level group of LBD approaches. Scientific knowledge can be naturally represented as a network, with nodes representing knowledge entities and edges/arcs representing the relationships between them. In this section, we use the term network strictly in the sense of a graph as a mathematical structure augmented with data (e.g., attributes on nodes and edges). To our knowledge, Wilkowski et al. (2011) were the first who made a significant contribution to this direction of LBD research. They represented semantic relations extracted with SemRep as a complex network. The arguments of the predicates (i.e. object and subject) were represented as nodes, while the predicates were mapped to arcs (i.e. directed edges). The idea was to extend the basic ABC model by the so-called AnC model, where the intermediate B element expands to form a walk, formally: $a - (b_1 - b_2 - \cdots - b_n) - c$.

While in the previously discussed ABC model, a discovery process consists of just two hops (i.e. $a - b$ and $b - c$), in the AnC model, the discovery process resembles an exploratory traversal of the network from node a to node c through a (set of) walk(s). This paradigm is also sometimes referred to as *discovery browsing*. As an example, this paradigm coupled with the Semantic MEDLINE Web tool was used to generate plausible explanations for *obesity paradox* (Cairelli et al., 2013).

Although machine learning in networks became popular in the mid-2000s, it took almost ten years for the first experiments to be conducted in the LBD community. *Link prediction* (or knowledge graph completion) refers to the discovery of future connections between nodes that are not directly connected in the current snapshot of a given network. Based on the principles of structural variation theory (Chen et al., 2009), it has been observed that the emergence of novel citation links between two distinct sets of articles often serves as a predictor of future scientific discoveries. This idea perfectly coincides with the notion of LBD. Therefore, one can view LBD as a link prediction problem among previously disjoint clusters of scientific papers.

Note that most of the data sources used in LBD (e.g., words, semantic relations, MeSH terms) could be represented and modeled more naturally as heterogeneous networks, which contain multiple types of nodes and edges. The heterogeneous network provides a new paradigm for managing relational data. Instead of relying only on lexical information (e.g., MeSH co-occurrence), the heterogeneous network approach also provides non-lexical information (e.g., author and citation relations). In this vein, Kostoff (2014) demonstrated that certain intermediate concepts that connect disjoint Parkinson's and Crohn's disease papers could only be found in the titles of their shared references. In this monograph, a selected network-based link prediction LBD approach is presented in Chapter 8.

Hybrid approaches. To comply with the categorization proposed by Gopalakrishnan et al. (2019), we list other approaches into the 'hybrid' category, including the unconventional open discovery approaches.

Ohsawa and Tsumoto (2006) demonstrate that tools that indicate rare events or situations play a significant role in research and discovery processes. From this perspective, researchers should be sensitive to curious or rare observations of phenomena, which indicate possible new opportunities for reasoning (Magnani, 2005) and be aware of the powerful support that data mining tools can have when choosing meaningful scenarios (Ohsawa & Tsumoto, 2006).

Outliers have attracted a lot of attention in the research world and are becoming increasingly popular in text mining applications as well. Detecting interesting outliers that rarely appear in a text collection can be viewed as the search for a needle in a haystack. This popular phrase illustrates the problem with rarity and outliers, as identifying useful rare objects is a difficult task itself (Magnani, 2005).

The rarity principle was applied in the RaJoLink literature-based open discovery process. In our earlier work (Petrič et al., 2007, 2009; Urbančič et al., 2007), we presented an approach that extends Swanson's *ABC model* to handle the open discovery process with rare terms from the domain literature. In this monograph, selected hybrid approaches are presented in Section 6.5 of Chapter 6, and Section 7.3 of Chapter 7.

Embeddings- and LLM-based approaches. Note that if one intends to employ machine learning techniques on graphs, it is required to learn informative network representations. Representation learning (Lavrač et al., 2021) provides an attractive alternative to conventional feature engineering. The main difficulty in machine learning on networks is finding a mechanism to include information about the structure of the network in the statistical model. For example, in link prediction, one aims to encode pairwise attributes between nodes, such as the number of shared neighbors or relationship strength. Traditional approaches rely on summary network statistics (e.g., centrality metrics) to extract structural information from networks. On the other hand, representation techniques aim to encode structural information about a network or its components (i.e. nodes and edges) in a high-dimensional vector space. The goal of graph embedding algorithms is to learn a mapping that embeds nodes in a multidimensional vector space by optimizing this mapping so that geometric relationships in this space match the structure of the original network. The learned embeddings can subsequently be used as automatically generated features for downstream machine learning tasks.

Various knowledge graph embedding techniques have been presented in the literature. Commonly used approaches include translational models (e.g., TransE, TransH, RotatE) that incorporate distance-based scoring functions and semantic matching models (e.g., RESCAL, DistMult, ComplEx), which utilize similarity-based scoring functions. Graph convolutional networks as well as contextualized encoding approaches have also been proposed. Note that for LBD, knowledge graph embeddings have been created from SemMedDB and used for link prediction for tasks such as drug discovery and drug repurposing.

Embeddings-based and Large Language Models (LLM)-based approaches form a separate category in the categorization that we used in structuring this monograph. In this monograph, an embeddings-based and an LLM-based approach are presented in Chapter 9 and Chapter 10), respectively.

2.2 Summary of LBD review papers

LBD is a complex, continually evolving, and collaborative research field. In recent years, the volume of scientific publications in the field of LBD has increased. Before reading the most important original papers, we encourage the reader who desires to get a deeper perspective on the field to start with recent literature surveys. Until now, at least a dozen literature surveys have been published to elucidate the extent of knowledge in LBD research. The LBD approaches, methods and applications have been systematically reviewed and classified by several authors, including the work of Ahmed (2016), Sebastian et al. (2017a), Henry and McInnes (2017), Gopalakrishnan et al. (2019), Thilakaratne et al. (2019a), and Thilakaratne et al. (2019b). A summary of these surveys, largely based on Kastrin and Hristovski (2021), is presented below.

Only three years after Swanson's first empirical paper on LBD, Davies (1989) published an interesting theoretical paper on the creation of new knowledge by information retrieval. This article provides an in-depth review of previous work on generating knowledge through information retrieval and presents methods to retrieve latent knowledge from the literature (e.g., serendipity browsing, proper search strategies, relational indexing, and artificial intelligence). Together with some more recent theoretical papers (Chen et al., 2009; Uzzi et al., 2013), the paper by Davies (1989) is a must-read for all researchers who aim to understand the formal foundations of LBD.

With the rise of various technologies (e.g., JavaScript) in the mid-1990s that stimulated the development of interactive Web applications, the LBD community started to build online LBD tools and services. Weeber et al. (2005) provided a review of methodology and LBD tools that had been developed until 2005. The authors described Arrowsmith (Swanson & Smalheiser, 1997), BITOLA (Hristovski et al., 2005), Manjal (Srinivasan, 2004), LitLinker (Pratt & Yetisgen-Yildiz, 2003), ACS (van der Eijk et al., 2004), Iridescent (Wren et al., 2004), and Telemakus (Fuller et al., 2004).

Bekhuis (2006) described LBD in the context of conceptual biology and the broader domain of text mining. The author provided a general background for knowledge discovery, a brief review of Swanson's ideas, and a short discussion of approaches for hypothesis discovery. Her review is complementary to the overviews published by Cohen (2005) and Jensen et al. (2006) around the same time.

Smalheiser wrote two review papers on LBD. In the first article, Smalheiser (2012) provided a critical overview of the then-prevalent ABC model. He concluded that the ABC paradigm was not wrong, however, it was only one of the many approaches to LBD that could stimulate the development of a new generation of LBD tools. Moreover, he advocated that we urgently needed some sort of objective function (i.e. interestingness measures) for filtering out the output of LBD systems. The second review by Smalheiser (2017) was written from a more personal perspective. The author discussed Swanson's contributions to LBD and gave an outline of future directions of LBD research.

Ahmed (2016) provided the first attempt to systematically classify LBD methods and approaches exclusively through the lens of information retrieval. Nearly at the same time, Sebastian et al. (2017a) and Henry and McInnes (2017), published extensive papers on LBD. The first group of authors provided an in-depth discussion on a broad palette of existing LBD approaches and offered performance evaluations on some emerging LBD methodologies. The latter authors likewise introduced historical and modern LBD approaches and provided an overview of evaluation methodologies and current trends. Both papers provided a general unifying framework for the LBD paradigms, methodologies, and tools.

The next review was published by Gopalakrishnan et al. (2019). Their paper provided a comprehensive analysis of the LBD field, and the paper served as a methodological introduction to particular techniques and tools. The authors provided a detailed discussion of the key LBD systems through the critical analysis of selected influential papers. They also summarized more recent research trends and identified future directions of LBD. Thilakaratne et al. (2019a) analyzed the methodologies used in LBD using a novel classification scheme (i.e. the main points of the review were computational techniques, central research topics, available tools, and applications) and provided a timeline with key milestones in LBD research. The authors also identified the current trends in LBD regarding publication over the years, top cited papers, and top authors. However, they considered only journals and conference proceedings for the review. In their second paper, Thilakaratne et al. (2019b) presented a large-scale systematic review of the LBD workflow by manually analyzing 176 LBD papers. Although these reviews successfully provide qualitative insight into the field of LBD by dissecting the research evidence and appropriate classification of research themes, their analysis was manual and did not offer quantitative examination.

2.3 LBD research topics and keywords

Although the aforementioned reviews are quite timely, an in-depth quantitative analysis in the LBD research field is urgently needed, to provide newcomers and also senior researchers with a state-of-the-art bibliometric and scientometric overview of the field. It is also important to keep researchers informed about emerging trends and essential turning points in the expansion of domain knowledge. Chen and Song (2019) performed the first knowledge mapping of the LBD field. However, the authors used the LBD domain as a proxy to illustrate an intuitive method to compare multiple search strategies in order to identify the most representative body of scientific publications. More recently, Kastrin and Hristovski (2021) presented the first comprehensive scientometric overview of LBD research, outlined in Section 2.3.2. The primary goal of the analysis was to examine and summarize a knowledge structure and evolution of LBD literature.

To provide further insights into the LBD field, the next subsections present a longitudinal analysis of the topics of LBD papers, world cloud summarization of the field, longitudinal keywords analysis, and a longitudinal analysis of papers when categorized into two simple categories 'methodology' and 'application' only.

2.3.1 Word cloud visualization of the field

The main topics studied in the LBD research field can be roughly characterized by the keywords assigned to domain scientific papers. The keywords extracted from scientific papers on LBD research provide a plausible summarization of research hotspots in the field. The word cloud presentation of the field shown in Figure 2.1 (Kastrin & Hristovski, 2021) enables us to summarize, qualify, and explain the given corpus of scientific documents created by LBD researchers. The visualization has been generated from a corpus comprised of 409 papers in the period of 35 years (1986–2020), beginning with Swanson's first publication on fish oil and Raynaud's disease. We utilized the wordcloud[3] package, with the size of each word or phrase proportional to its frequency in the corpus. For more details, refer to Kastrin and Hristovski (2021).

2.3.2 Analysis of topics through a scientometric lens

Kastrin and Hristovski (2021) utilized a comprehensive scientometric approach to systematically analyze the LBD literature published in the last four decades (1986–2020). Document co-citation analysis (DCA) allowed them to examine a network of co-cited references. The body of cited references provides the knowledge base of the selected documents. Figure 2.2 depicts the subset of the DCA network as pruned with the Pathfinder algorithm (Chen, 2006). The network exhibits 399 nodes and 891 edges. Each node in the network refers to a document that is labeled with the first author's name and the year of publication. Each edge represents a co-citation relation among the pair of documents. The size of the nodes is proportional to the co-citation frequency. The interested reader can find the description of the clusters in the paper by Kastrin and Hristovski (2021)

[3] https://github.com/amueller/word_cloud

Fig. 2.1 Word cloud of keywords extracted from publications on LBD research (Kastrin & Hristovski, 2021). Note that terms *literature-based discovery* and LBD have been removed from the plot.

2.3.3 Longitudinal keywords analysis

The research until 2000 was topical and domain-specific, loaded with keywords such as *fish oil*, *information retrieval*, *raynaud*, *magnesium*, and *migraine*. In the new millennium, the richness of keywords increased, and the terms rapidly evolved. In the 2010s, the authors were able to observe two main research directions: one concerned with interweaving LBD ideas with genetics, and another concerned with the application of semantics in LBD. The most prominent keywords were *text mining*, *knowledge discovery*, *data mining*, *disease*, *gene*, and *natural language processing*. This indicates that LBD research in this decade was developing from the baseline Swanson's approach and its replications (Gordon & Lindsay, 1996) towards systematic development of knowledge discovery methods and data mining tools for LBD. In terms of LBD reviews from this era, the authors observed a shift from statistical approaches (e.g., pure co-occurrence) to rule-based (e.g., association rules) and semantic approaches. In this decade, numerous other Web tools and services were also developed (e.g., DAD (Weeber et al., 2001), LitLinker (Pratt & Yetisgen-Yildiz, 2003), Manjal (Srinivasan & Libbus, 2004), Iridescent (Wren et al., 2004), and RaJoLink (Petrič et al., 2009)). The next decade of LBD (i.e. 2011–2020) was the decade of network science in the LBD community. The important keywords in this time period were *network*, *link discovery*, *link prediction*, *network analysis*, *knowledge graph*, *drug discovery*, and *pharmacovigilance*. According to Sebastian et al. (2017a) this decade coincided with the stage of emerging LBD approaches which could be characterized by two directions. First, traditional co-occurrence-based and knowledge-driven approaches culminated in solutions that integrate both. Second, LBD became more interdisciplinary, incorporating methods

Fig. 2.2 Document co-citation network of the core dataset based on documents published between 1986 and 2020, published by Kastrin and Hristovski (2021). Cluster labels in red text are taken from the titles of the cited documents using the log-likelihood ratio algorithm. Red nodes refer to documents with high citation bursts.

and tools from information sciences (Chen et al., 2009), scientometrics (Kostoff, 2014), and machine learning (Sebastian et al., 2017b).

2.3.4 Longitudinal analysis of two selected topics

For this analysis, performed by Koloski et al. (2024), we have compiled a novel comprehensive corpus of LBD publications by merging lists of representative publications

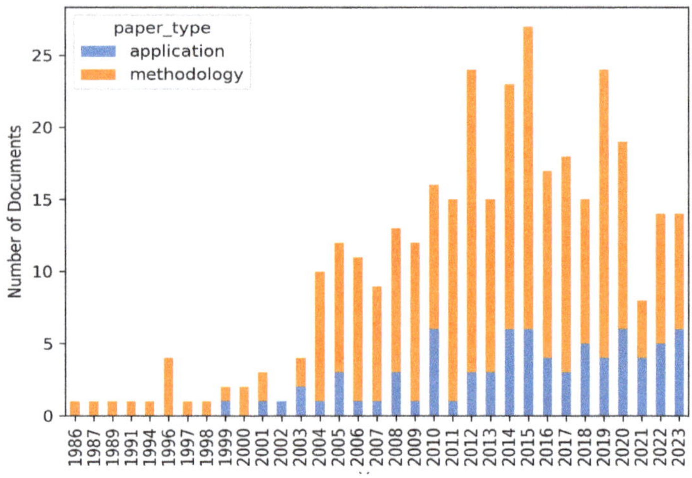

Fig. 2.3 Documents distribution per year, semantically labeled by LLaMa2 prompting.

on LBD that were manually compiled by the authors of two recent surveys of the LBD field (Kastrin & Hristovski, 2021; Thilakaratne et al., 2019b). To ensure that the latest advances were included, we also added 11 papers published in 2022 and 2023 which were not included in the previous surveys. The corpus analyzed contained 389 publications published from 1986 to 2023. We concatenated the title and abstract fields to eliminate empty features because six papers did not contain an abstract.

This analysis was performed using a Large Language Model (LLM) approach. We prompted the large language model LLaMa2 (Touvron et al., 2023), briefly introduced in Section 3.6, to semantically categorize LBD articles into two groups:

- *methodology* articles, mainly proposing a new approach or methodology, and
- *application* articles, mainly applying some already developed methods.

Our goal in this categorization task was to trace the evolution of the field, assuming that the 'methodology' articles are those that introduce new techniques, while the 'application' articles are likely to focus on the use or refinement of existing methods. We report the results in Figure 2.3.

The first LBD paper, published in 1986, reporting Swanson's discovery of a link between fish oil and Raynaud's disease (Swanson, 1986a), was categorized by the LLM as a 'methodology' paper. The longitudinal analysis shows that in the early years, LBD literature mostly explored methodological issues. Over time, the focus of the work expanded to include both 'methodology' and 'application', with the latter gaining traction from 1999 onward. A notable surge in 'methodology' articles occurred around 2008, with a peak in 2015, followed by some variations over the years. On the other hand, 'application' articles saw a steady increase, underscoring the sustained interest and necessity for LBD research. By 2023, the production of both 'methodology' and 'application' articles appears to have stabilized, indicating a balanced advancement in these two areas.

2.4 Standard LBD resources

With the advent of internet, scientific communication and the volume of work made easily available to researchers has grown exponentially, and with it new possibilities for finding interesting information. This section describes the selected public literature resources in Section 2.4.1, followed by the description of semantic LBD resources in Section 2.4.2, while the description of benchmark data sets that can be used to test the utility of LBD tools can be found in Section 4.1.

2.4.1 Public literature resources

The availability of online databases such as MEDLINE, accessible through PubMed, as a rich source of information was an important factor for literature mining applications to be focused mostly on biomedical domains, although the literature mining methods are general and could be used in other domains using other sources.

MEDLINE. Being the largest PubMed resource, MEDLINE is a database containing over 30 million biomedical and life sciences journal references, compiled and maintained by the US National Institute of Health, National Library of Medicine. The overall 2020 MEDLINE baseline encompasses approximately 5,600 journals and contains 32,775,609 references, of which 23,835,206 contain abstracts. The approaches described in this monograph mainly use MEDLINE titles and abstracts from 1975 to the present day. Before 1975, only 2% of the citations contained an abstract. The percentage of publications that include abstracts steadily increased over time. Figure 2.4 shows the volume of MEDLINE records over time. The database gets an update from Monday through Saturday, which usually brings in from 2,000 to 4,000 new records a day. The MEDLINE distribution is available in a machine-readable format and the entire distribution may be downloaded for further processing.

PubMed. PubMed is a freely available search engine, which includes bibliographic data for papers in most academic journals in the life sciences, covering all topics in medicine and biology and their many subfields. These account for virtually all papers published in the English language (in the life sciences) since the mid-20th century, as well as a small percentage of papers published in other languages. In many cases, full texts of papers are indexed as well, though these depend on the journal and, in the case of papers behind a paywall, the subscription status of the user or of the institution providing the connection to the database. In addition to MEDLINE, PubMed service contains PubMed Central (a free repository of more than 7.5 million full-text papers in a machine-readable format and richly associated

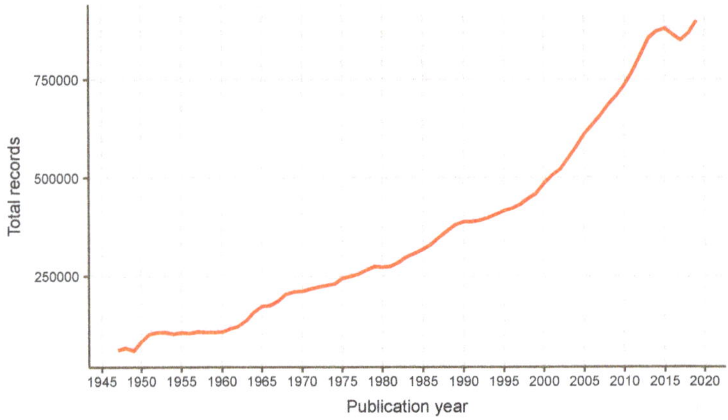

Fig. 2.4 MEDLINE citation counts.

with metadata), NCBI Bookshelf (a vast selection of freely available eBooks on life sciences), and other paper repositories. While PubMed Central contains freely available full-text papers, MEDLINE contains just the references composed of a title, author information, and in about 65% of the references also an abstract.

2.4.2 Semantic LBD resources

Given the quick and easy access to virtually all relevant publications that an LBD researcher could think of using PubMed resources, one of the main challenges for the researcher is the size of the search space, which can be huge. This problem is addressed in different ways by different approaches, e.g., by using Medical Subject Headings (MeSH) concepts or Unified Medical Language System (UMLS) semantic types as filters. This is especially useful in the case that a given system is specialized for a particular task, with a particular kind of data in mind. For example, MeSH can be used as a source of background knowledge to provide a thesaurus of gene and protein symbols (Hristovski et al., 2005).

Unified Medical Language System (UMLS). The basic unit of UMLS[4] is a concept (Bodenreider, 2004). The concept represents the term's meaning and also encompasses the syntactic and lexical variations of individual terms. Concepts are organized in the form of a directed acyclic graph that reflects their relationships. For example, in Figure 2.5, the node labeled "heart" refers to the concept with the broadest meaning, while the adjacent nodes represent concepts with more specific meanings.

UMLS is composed of three components:

[4] nlm.nih.gov/research/umls

(i) Metathesaurus, which includes over 250 vocabularies broadly connected to biomedicine (e.g., MeSH, SNOMED CT, Gene Ontology, OMIM). All variations of a term with a similar/unique meaning are mapped to a Concept Unique Identified (CUI). Each concept is assigned a Semantic Type (127 in total, as of this writing), which can be further mapped to a Semantic Group (15 in total).
(ii) SPECIALIST lexicon, which contains lexical data for frequent biomedical words.
(iii) Semantic Network, which describes the links among the Metathesaurus concepts. Formally, the Semantic Network is a heterogeneous graph with nodes referring to particular concepts in the Metathesaurus and (un)directed links referring to semantic relations between the network nodes.

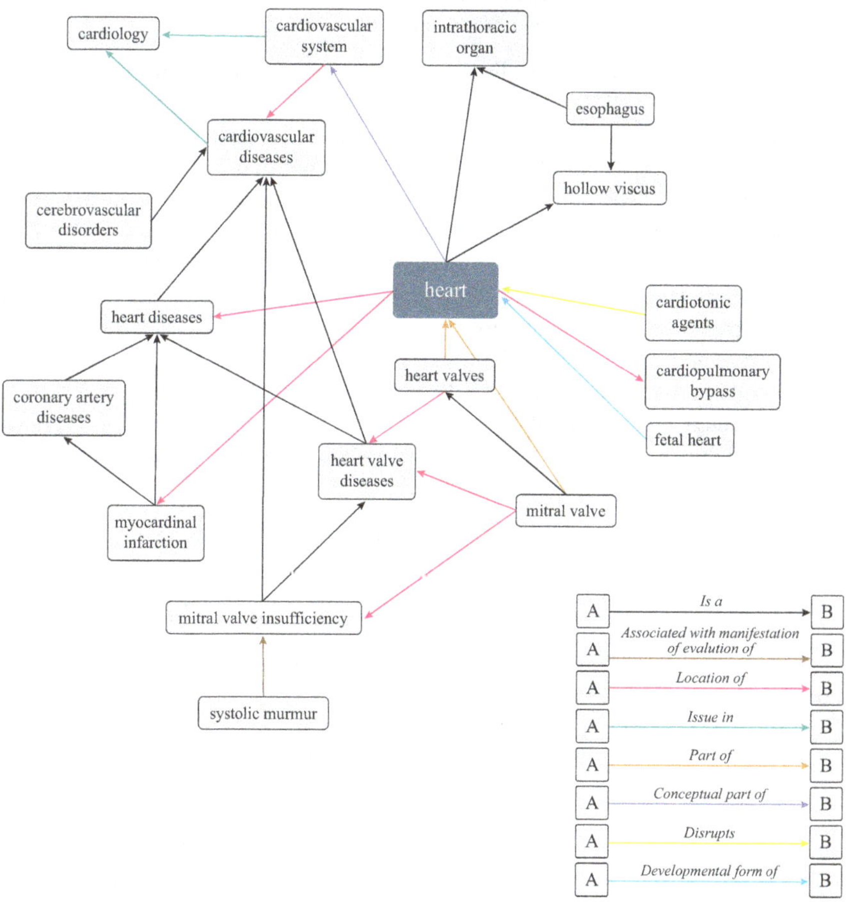

Fig. 2.5 Partial representation of the concept *heart* in UMLS.

The UMLS Semantic Network[5] comprises two components: (i) a set of 134 broad subject categories, known as Semantic Types, which ensure a categorization of knowledge concepts within the UMLS Metathesaurus; and (ii) a set of 54 relationships, referred to as Semantic Relations, that connect pairs of Semantic Types. The UMLS Metathesaurus assigns one or more semantic types to every concept, ensuring a consistent and source vocabulary-independent categorization in the network. Semantic types are arranged in a hierarchy that is organized into two main categories, i.e. *Entity* and *Event*. Examples of the former semantic types are '*Gene or Genome* and *Carbohydrate*, while representatives of the latter semantic types include categories *Laboratory Procedure* and *Mental Process*.
Large-scale semantic taxonomies, such as the UMLS, allow for the identification of medical terminology in text. Existing tools, such as MetaMap and cTAKES, provide programmatic means for mapping text to concepts in the UMLS.
The basic mapping strategy is composed of the following steps:

1. Parse the text into noun phrases and perform the remaining steps for each phrase;
2. Generate the variants for the noun phrase where a variant essentially consists of one or more noun phrase words together with all of its spelling variants, abbreviations, acronyms, synonyms, inflectional and derivational variants, and meaningful combinations of these;
3. Form the candidate set of all Meta strings containing one of the variants;
4. For each candidate, compute the mapping from the noun phrase and calculate the strength of the mapping using an evaluation function. Order the candidates by mapping strength; and
5. Combine candidates involved with disjoint parts of the noun phrase, recompute the match strength based on the combined candidates, and select those having the highest score to form a set of best Meta mappings for the original noun phrase.

Medical Subject Headings (MeSH). MeSH[6] is a controlled vocabulary (i.e. a thesaurus) used primarily for indexing biomedical literature and to support searching in library catalogs (Rogers, 1963). MeSH provides uniform and consistent labeling of bibliographic records indexed by the US National Library of Medicine in MEDLINE. MeSH is hierarchical and includes several types of terms: *Main Headings* (or descriptors), *Subheadings* (also called qualifiers), *Supplementary Concept Records*, and *Publication Characteristics*. Main headings represent knowledge concepts that occur in the free text. For example, in the sentence

> The main objective was to examine prescription patterns of glucose-lowering therapy among adults with PPDM compared with type 1 and type 2 diabetes [PMID: 34362812].

the phrases *glucose-lowering therapy* and *type 2 diabetes* are mapped to the main headings *Hypoglycemic Agents* and *Diabetes Mellitus, Type 2*, respectively. MeSH subheadings are used to expand the meaning of the descriptor by describing a

[5] https://lhncbc.nlm.nih.gov/semanticnetwork/

[6] nlm.nih.gov/mesh

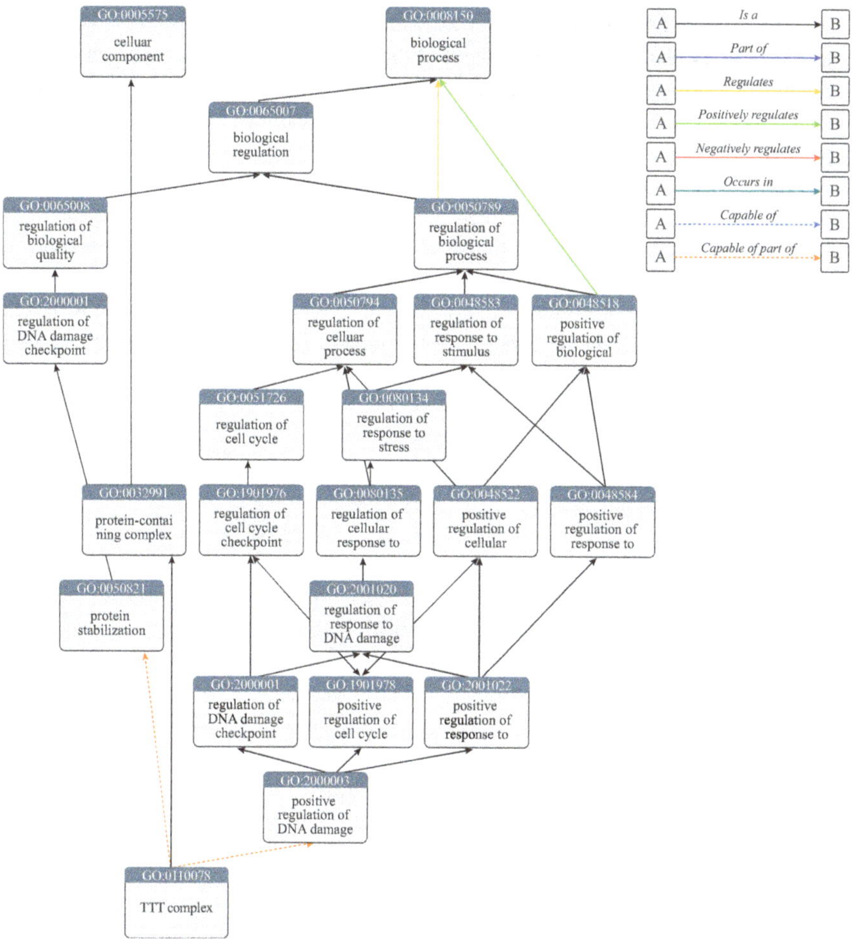

Fig. 2.6 A part of the Gene Ontology (Ashburner et al., 2000) related to cell regulation, with multiple types of relations (directed edges of different colors), demonstrating that ontologies include hierarchical relations.

thematic background of a particular knowledge concept. In the example above, the possible subheading for the descriptor *Diabetes Mellitus, Type 2* would be *drug therapy* that better guides the reader towards diabetes therapy.

Gene Ontology (GO). With the introduction of the semantic web (Berners-Lee et al., 2001), ontologies emerged as the standard data structure for working with background knowledge. In computational biology, examples of structured domain knowledge include the Gene Ontology (GO) (Ashburner et al., 2000), the KEGG ontology (Ogata et al., 1999) and the Entrez gene–gene interaction data (Maglott et al., 2005). As an illustrative example, widely used in bioinformatics, take the well-known Gene Ontology (Ashburner et al., 2000), which contains tens of thousands of interconnected biological terms, corresponding to cellular and functional processes between individual genes, as well as groups of genes described by higher-level ontology terms. A small segment of the Gene Ontology is shown in Figure 2.6. Formally, ontologies are directed acyclic graphs, formed of concepts and their relations, encoded as *subject-predicate-object* (s, p, o) triplets. The simplest ontologies are *taxonomies*, where nodes are hierarchically ordered from more general nodes at higher levels of the hierarchy to more specific concepts at lower levels of the hierarchy. Nodes are connected by *is-a* or *part-of* relationships between the nodes. Ontologies can include other (non-hierarchical) types of relations and represent richer data structures than taxonomies. In both ontologies and taxonomies, bottom-level nodes describe individual instances (e.g., individual genes in the GO), while higher-level nodes describe concepts or groups of instances (e.g., groups of genes in the GO).

2.5 Selected LBD tools

In their review of methodology and LBD tools that had been developed until 2005, Weeber et al. (2005) described several tools and services, including Arrowsmith (Swanson & Smalheiser, 1997), BITOLA (Hristovski et al., 2005), Manjal (Srinivasan, 2004), LitLinker (Pratt & Yetisgen-Yildiz, 2003), ACS (van der Eijk et al., 2004), Iridescent (Wren et al., 2004), and Telemakus (Fuller et al., 2004). To our knowledge, only the Arrowsmith tool is still regularly updated, improved, and available to the broad research public.

This section first presents a selection of data acquisition and preprocessing tools in Section 2.5.1, followed by selected LBD support tools in Section 2.5.2.

2.5.1 Selected document acquisition tools

This section presents selected tools for document acquisition and processing.

Entrez Programming Utilities. Entrez is a specialized database system produced by the National Center for Biotechnology Information (NCBI) that provides access to nucleotide and protein data, genomic mapping information, PubMed, and a lot more. To enable programmatic access to the data, NCBI provides a tool called E-utilities, which consists of nine programs listed in Table 2.1.

MetaMap. MetaMap is an NCBI tool for normalizing biomedical free text to UMLS concepts and semantic relations. An input to the MetaMap is a raw text (usually the title and/or abstract of a paper), and an output is a sequence of concepts and semantic types.

SemRep. SemRep was proposed by Rindflesch and Fiszman (2003) to fill the gap between raw (biomedical) text and its meaning. SemRep is a NLP system, which extracts semantic relationships among biomedical concepts in free text. SemRep is based on a partial syntactic analysis and structured domain knowledge from the UMLS. Semantic relationships are represented as predications consisting of a subject, predicate, and object. Subject and object arguments are concepts from the UMLS Metathesaurus as accessed through MetaMap. UMLS Semantic Network links serve as predicates. Predications provide a normalized form of the meaning of the free text. As an example, SemRep extracts the predication `Genes AFFECTS Circadian Rhythms` from the sentence "Clock genes are the genes that control circadian rhythms in physiology and behavior".

SemRep extracts various predications related to molecular interactions (e.g., `INTERACTS_WITH`, `INHIBITS`, `STIMULATES`), disease etiology (e.g., `ASSOCIATED_WITH`, `CAUSES`, `PREDISPOSES`), clinical medicine (e.g., `DIAGNOSES`, `PROCESS_OF`, `TREATS`), or pharmacogenomics (e.g., `AFFECTS`, `AUGMENTS`, `DISRUPTS`).

SemMedDB. SemMedDB was proposed by Kilicoglu et al. (2012) who processed the entire MEDLINE baseline distribution with the SemRep NLP tool and extracted more than 23 million citations and about 68 million semantic predications. Predications were stored in a MySQL database and made freely available to the research community. SemMedDB is now updated biannually (as of this writing, SemMedDB contains about 110 million predications from over 35 million MEDLINE/PubMed bibliographic records.

Table 2.1 List of programs that provide access to the NCBI Entrez databases.

Utility	Main functionality
EInfo	Provides the number of records indexed in each field of a given database, the date of the last update of the database, and the available links from the database to other Entrez databases.
ESearch	Responds to a text query with the list of matching UIDs in a given database (for later use in ESummary, EFetch or ELink), along with the term translations of the query.
EPost	Accepts a list of UIDs from a given database, stores the set on the History Server, and responds with a query key and web environment for the uploaded dataset.
ESummary	Responds to a list of UIDs from a given database with the corresponding document summaries.
EFetch	Responds to a list of UIDs in a given database with the corresponding data records in a specified format.
ELink	Responds to a list of UIDs in a given database with either a list of related UIDs (and relevancy scores) in the same database or a list of linked UIDs in another Entrez database.
EGQuery	Responds to a text query with the number of records matching the query in each Entrez database.
ESpell	Retrieves spelling suggestions for a text query in a given database.
ECitMatch	Retrieves PMIDs corresponding to a set of input citation strings.

2.5.2 Selected LBD support tools

This section presents brief descriptions of selected LBD support tools applied in some of the approaches presented in this monograph. Table 2.2 presents a collection of computational tools developed by the LBD community.

Table 2.2 LBD tools with key technologies.

Tool	Reference and Technology Description
Arrowsmith	Smalheiser and Swanson (1998) Statistical (relative and absolute frequencies), knowledge-based (MeSH, UMLS), machine learning (aggregate local and global features to obtain a composite ranking function)
LitLinker	Pratt and Yetisgen-Yildiz (2003) Probabilistic (z-score), knowledge-based (MeSH), association rule mining (support and confidence)
RaJoLink	Petrič et al. (2009) Statistical (rarity), knowledge-based (MeSH, HUGO, ToppGene, Endeavour, STRING)
Bitola	Hristovski et al. (2005) Knowledge-based (UMLS), predicate-based (SemRep, BioMedLEE), discovery patterns (two disease treatment patterns)
Manjal	Srinivasan (2004) Statistical (TF-IDF), knowledge-based (MeSH, UMLS), profile-based (weighted vectors with semantic type groupings)
DAD	Weeber et al. (2000) Statistical (frequency), knowledge-based (UMLS)
Anni	Jelier et al. (2008) Knowledge-based (UMLS, Gene Ontology), profile-based (weighted related concepts)
Iridescent	Wren et al. (2004) Statistical (mutual information measure), knowledge-based (MeSH, OMIM, drug names from FDA, LocusLink, Gene Ontology)
CrossBee	Juršič et al. (2012) Statistical (elementary and ensemble heuristics for term ranking)
LION LBD	Pyysalo et al. (2019) Knowledge-based (PubTator), network-based path mining
AGATHA	Sybrandt et al. (2020) Transformer-based
MELODI Presto	Elsworth and Gaunt (2021) Knowledge-based (SemMedDB)

References

Ahmed, A. (2016). Literature-based discovery: Critical analysis and future directions. *International Journal of Computer Science and Network Security*, *16*(7), 11–26.

Ashburner, M., Ball, C. A., Blake, J. A., Botstein, D., Butler, H., Cherry, J. M., Davis, A. P., Dolinski, K., Dwight, S. S., Eppig, J. T., et al. (2000). Gene ontology: Tool for the unification of biology. *Nature Genetics*, *25*(1), 25–29. https://doi.org/10.1038/75556

Bekhuis, T. (2006). Conceptual biology, hypothesis discovery, and text mining: Swanson's legacy. *Biomedical Digital Libraries*, *3*(1). https://doi.org/10.1186/1742-5581-3-2

Berners-Lee, T., Hendler, J., & Lassila, O. (2001). The semantic web. *Scientific American*, *284*(5), 34–43.

Bodenreider, O. (2004). The Unified Medical Language System (UMLS): Integrating biomedical terminology. *Nucleic Acids Research*, *32*(Database issue), D267–270. https://doi.org/10.1093/nar/gkh061

Cairelli, M. J., Miller, C. M., Fiszman, M., Workman, T. E., & Rindflesch, T. C. (2013). Semantic MEDLINE for discovery browsing: Using semantic predications and the literature-based discovery paradigm to elucidate a mechanism for the obesity paradox. *AMIA Annual Symposium Proceedings*, *2013*, 164.

Chen, C. (2006). CiteSpace II: Detecting and visualizing emerging trends and transient patterns in scientific literature. *Journal of the American Society for Information Science and Technology*, *57*(3), 359–377. https://doi.org/10.1002/asi.20317

Chen, C., Chen, Y., Horowitz, M., Hou, H., Liu, Z., & Pellegrino, D. (2009). Towards an explanatory and computational theory of scientific discovery. *Journal of Informetrics*, *3*(3), 191–209.

Chen, C., & Song, M. (2019). Visualizing a field of research: A methodology of systematic scientometric reviews. *PLOS ONE*, *14*(10), e0223994. https://doi.org/10.1371/journal.pone.0223994

Cohen, A. M. (2005). A survey of current work in biomedical text mining. *Briefings in Bioinformatics*, *6*(1), 57–71. https://doi.org/10.1093/bib/6.1.57

Davies, R. (1989). The creation of new knowledge by information retrieval and classification. *Journal of Documentation*, *45*(4), 273–301. https://doi.org/10.1108/eb026846

Elsworth, B., & Gaunt, T. R. (2021). Melodi presto: A fast and agile tool to explore semantic triples derived from biomedical literature. *Bioinformatics*, *37*(4), 583–585.

Fine, G. A., & Deegan, J. G. (1996). Three principles of Serendip: Insight, chance, and discovery in qualitative research. *International Journal of Qualitative Studies in Education*, *9*(4), 434–447.

Fuller, S. S., Revere, D., Bugni, P. F., & Martin, G. M. (2004). A knowledgebase system to enhance scientific discovery: Telemakus. *Biomedical Digital Libraries*, *1*(1), 2.

García, P. (2009). Discovery by serendipity: A new context for an old riddle. *Foundations of Chemistry*, *11*(1), 33–42.

Gopalakrishnan, V., Jha, K., Jin, W., & Zhang, A. (2019). A survey on literature based discovery approaches in biomedical domain. *Journal of Biomedical Informatics*, *93*, 103141.

Gordon, M. D., & Lindsay, R. K. (1996). Toward discovery support systems: A replication, re-examination, and extension of Swanson's work on literature-based discovery of a connection between Raynaud's and fish oil. *Journal of the American Society for Information Science*, *47*(2), 116–128.

Henry, S., & McInnes, B. T. (2017). Literature based discovery: Models, methods, and trends. *Journal of Biomedical Informatics*, *74*, 20–32.

Hristovski, D., Peterlin, B., Mitchell, J. A., & Humphrey, S. M. (2005). Using literature-based discovery to identify disease candidate genes. *International Journal of Medical Informatics*, *74*(2), 289–298.

Jelier, R., Schuemie, M. J., Veldhoven, A., Dorssers, L. C., Jenster, G., & Kors, J. A. (2008). Anni 2.0: A multipurpose text-mining tool for the life sciences. *Genome Biology*, *9*(6), R96.

Jensen, L. J., Saric, J., & Bork, P. (2006). Literature mining for the biologist: From information retrieval to biological discovery. *Nature Reviews Genetics*, *7*(2), 119–129.

Juršič, M., Cestnik, B., Urbančič, T., & Lavrač, N. (2012). Cross-domain literature mining: Finding bridging concepts with CrossBee. *Proceedings of the 3rd International Conference on Computational Creativity*, 33–40.

Kastrin, A., & Hristovski, D. (2021). Scientometric analysis and knowledge mapping of literature-based discovery (1986–2020). *Scientometrics*, *126*(2), 1415–1451.

Kilicoglu, H., Rosemblat, G., Fiszman, M., & Shin, D. (2020). Broad-coverage biomedical relation extraction with SemRep. *BMC Bioinformatics*, *21*(1), 188.

Kilicoglu, H., Shin, D., Fiszman, M., Rosemblat, G., & Rindflesch, T. C. (2012). SemMedDB: A PubMed-scale repository of biomedical semantic predications. *Bioinformatics*, *28*(23), 3158–3160.

Koloski, B., Lavrač, N., Cestnik, B., Pollak, S., Škrlj, B., & Kastrin, A. (2024). AHAM: Adapt, Help, Ask, Model Harvesting LLMs for literature mining. *Advances in Intelligent Data Analysis XXII. IDA 2024.*, 254–265.

Kostoff, R. N. (2014). Literature-related discovery: Common factors for Parkinson's disease and Crohn's disease. *Scientometrics*, *100*(3), 623–657.

Lavrač, N., Podpečan, V., & Robnik-Šikonja, M. (2021). *Representation learning: Propositionalization and embeddings*. Springer.

Maglott, D., Ostell, J., Pruitt, K. D., & Tatusova, T. (2005). Entrez Gene: Gene-centered information at NCBI. *Nucleic Acids Research*, *33*, D54–D58.

Magnani, L. (2005). Chance discovery and the disembodiment of mind. *International Conference on Knowledge-Based and Intelligent Information and Engineering Systems*, 547–553.

Nelson, S. J., Johnston, W. D., & Humphreys, B. L. (2001). Relationships in Medical Subject Headings (MeSH). In C. A. Bean & R. Green (Eds.), *Relationships in the organization of knowledge* (pp. 171–184). Springer.

Ogata, H., Goto, S., Sato, K., Fujibuchi, W., Bono, H., & Kanehisa, M. (1999). KEGG: Kyoto Encyclopedia of Genes and Genomes. *Nucleic Acids Research, 27*(1), 29–34.

Ohsawa, Y., & Tsumoto, S. (2006). *Chance Discoveries in Real World Decision Making: Data-based Interaction of Human Intelligence and Artificial Antelligence*. Springer.

Petrič, I., Urbančič, T., & Cestnik, B. (2007). Discovering hidden knowledge from biomedical literature. *Informatica, 31*(1), 15–20.

Petrič, I., Urbančič, T., Cestnik, B., & Macedoni-Lukšič, M. (2009). Literature mining method RaJoLink for uncovering relations between biomedical concepts. *Journal of Biomedical Informatics, 42*(2), 219–227.

Pratt, W., & Yetisgen-Yildiz, M. (2003). LitLinker: Capturing connections across the biomedical literature. *Proceedings of the 2nd International Conference on Knowledge Capture*, 105–112.

Pyysalo, S., Baker, S., Ali, I., Haselwimmer, S., Shah, T., Young, A., Guo, Y., Högberg, J., Stenius, U., Narita, M., & Korhonen, A. (2019). LION LBD: A literature-based discovery system for cancer biology. *Bioinformatics, 35*(9), 1553–1561.

Rindflesch, T. C., & Fiszman, M. (2003). The interaction of domain knowledge and linguistic structure in natural language processing: Interpreting hypernymic propositions in biomedical text. *Journal of Biomedical Informatics, 36*(6), 462–477.

Rogers, F. B. (1963). Medical Subject Headings. *Bulletin of the Medical Library Association, 51*, 114–116.

Sebastian, Y., Siew, E.-G., & Orimaye, S. O. (2017a). Emerging approaches in literature-based discovery: Techniques and performance review. *The Knowledge Engineering Review, 32*, e12.

Sebastian, Y., Siew, E.-G., & Orimaye, S. O. (2017b). Learning the heterogeneous bibliographic information network for literature-based discovery. *Knowledge-Based Systems, 115*, 66–79.

Smalheiser, N. R. (2012). Literature-based discovery: Beyond the ABCs. *Journal of the American Society for Information Science and Technology, 63*(2), 218–224.

Smalheiser, N. R. (2017). Rediscovering Don Swanson: The past, present and future of literature-based discovery. *Journal of Data and Information Science, 2*(4), 43–64.

Smalheiser, N. R., & Swanson, D. R. (1998). Using ARROWSMITH: A computer-assisted approach to formulating and assessing scientific hypotheses. *Computer Methods and Programs in Biomedicine, 57*(3), 149–154.

Srinivasan, P. (2004). Text mining: Generating hypotheses from MEDLINE. *Journal of the American Society for Information Science and Technology, 55*(5), 396–413.

Srinivasan, P., & Libbus, B. (2004). Mining MEDLINE for implicit links between dietary substances and diseases. *Bioinformatics, 20*(suppl_1), i290–i296.

Swanson, D. R. (1986a). Fish oil, Raynaud's syndrome, and undiscovered public knowledge. *Perspectives in Biology and Medicine, 30*(1), 7–18.

Swanson, D. R. (1986b). Undiscovered public knowledge. *The Library Quarterly, 56*(2), 103–118.

Swanson, D. R., & Smalheiser, N. R. (1997). An interactive system for finding complementary literatures: A stimulus to scientific discovery. *Artificial Intelligence, 91*(2), 183–203.

Sybrandt, J., Tyagin, I., Shtutman, M., & Safro, I. (2020). Agatha: Automatic graph mining and transformer based hypothesis generation approach. *Proceedings of the 29th ACM International Conference on Information & Knowledge Management*, 2757–2764.

Thilakaratne, M., Falkner, K., & Atapattu, T. (2019a). A systematic review on literature-based discovery: General overview, methodology, & statistical analysis. *ACM Computing Surveys, 52*(6), 129:1–129:34.

Thilakaratne, M., Falkner, K., & Atapattu, T. (2019b). A systematic review on literature-based discovery workflow. *PeerJ Computer Science, 5*, e235.

Touvron, H., Martin, L., Stone, K., Albert, P., Almahairi, A., & Babaei, Y. (2023). *Llama 2: Open Foundation and Fine-Tuned Chat Models.* ArXiv. https://doi.org/10.48550/arXiv.2307.09288

Urbančič, T., Petrič, I., Cestnik, B., & Macedoni-Lukšič, M. (2007). Literature mining: Towards better understanding of autism. *Proceedings of the 11th Conference on Artificial Intelligence in Medicine*, 217–226.

Uzzi, B., Mukherjee, S., Stringer, M., & Jones, B. (2013). Atypical combinations and scientific impact. *Science, 342*(6157), 468–472.

van der Eijk, C. C., van Mulligen, E. M., Kors, J. A., Mons, B., & van den Berg, J. (2004). Constructing an associative concept space for literature-based discovery. *Journal of the American Society for Information Science and Technology, 55*(5), 436–444.

Weeber, M., Klein, H., Aronson, A. R., Mork, J. G., de Jong-van den Berg, L. T., & Vos, R. (2000). Text-based discovery in biomedicine: The architecture of the DAD-system. *AMIA Symposium Proceedings*, 903–907.

Weeber, M. (2007). Drug discovery as an example of literature-based discovery. In S. Džeroski & L. Todorovski (Eds.), *Computational Discovery of Scientific Knowledge: Introduction, Techniques, and Applications in Environmental and Life Sciences* (pp. 290–306). Springer.

Weeber, M., Klein, H., de Jong-van den Berg, L., Vos, R., et al. (2001). Using concepts in literature-based discovery: Simulating Swanson's Raynaud–fish oil and migraine–magnesium discoveries. *Journal of the American Society for Information Science and Technology, 52*(7), 548–557.

Weeber, M., Kors, J. A., & Mons, B. (2005). Online tools to support literature-based discovery in the life sciences. *Briefings in Bioinformatics, 6*(3), 277–286.

Wilkowski, B., Fiszman, M., Miller, C. M., Hristovski, D., Arabandi, S., Rosemblat, G., & Rindflesch, T. C. (2011). Graph-based methods for discovery browsing with semantic predications. *Proceedings of the AMIA Annual Symposium, 2011*, 1514–1523.

Wren, J. D., Bekeredjian, R., Stewart, J. A., Shohet, R. V., & Garner, H. R. (2004). Knowledge discovery by automated identification and ranking of implicit relationships. *Bioinformatics, 20*(3), 389–398.

Yetisgen-Yildiz, M., & Pratt, W. (2006). Using statistical and knowledge-based approaches for literature-based discovery. *Journal of Biomedical Informatics, 39*(6), 600–611.

Chapter 3
Background Technologies

This chapter presents a selection of the basic technologies used as elementary components of LBD methods. Section 3.1 provides an outline of information extraction methods. Section 3.2 presents a standard natural language processing (NLP) pipeline used in text analysis. Section 3.3 presents the Bag-of-Words representation of text documents and similarity measures used in standard text mining tasks, such as clustering. Section 3.4 introduces the area of network analysis, Section 3.5 presents selected embedding-based analysis approaches, and Section 3.6 briefly introduces the Large Language Model (LLM) technology. The evaluation of performance of LBD systems and their text mining and network analysis ingredients are explained in Section 3.7.

3.1 Information extraction

Information extraction (IE) (Manning et al., 2008) refers to the process of automatically identifying and extracting structured information from unstructured or semi-structured machine-readable documents. This involves processing human language text using natural language processing. Typical tasks of IE include keyword extraction, named entity recognition, relation extraction, and document summarization. The targets for information extraction are terms, particularly keywords and named entities. While *terms*, defined as *verbal designations of general concepts in a specific subject field*, are most frequently assigned at a document collection level (i.e. for the entire domain), *keywords* and *named entities* are usually assigned at a single document level. The meaning of keywords and named entities are explained in the sequel.

Information extraction tasks that are the most relevant for this book include:

Keyword extraction (KE). Keywords are expressions that best describe the subject of a document and a good keyword effectively summarizes the content of a document allowing it to be efficiently retrieved when needed. The task of keyword extraction is to identify the most important terms in a document (Beliga et al., 2015). While not all terms are keywords, there is a strong overlap between the most frequent terms and keywords, therefore term extraction techniques can be applied successfully to

N. Lavrač et al., *Bisociative Literature-Based Discovery*,
https://doi.org/10.1007/978-3-031-96863-1_3

keyword extraction. Monolingual approaches, based on linguistics and statistics (word associations, comparison to monolingual dictionaries) or machine learning, can be applied to a large variety of tasks and texts.

Named entity recognition (NER). Named-entity recognition usually relies on language-dependent techniques to assign a unique identifier to an entity, such as a person's name, location, organization, etc. While named entities assume the role of keywords, named entity recognition is a distinct research area (Hoffart et al., 2011). For example, a geographic named entity such as *Brussels* found in an English text should be used as a representation across languages by a unique NE identifier also for texts in other languages, including e.g., the word *Bruxelles* in French. Similarly, in medical data analysis, Alzheimer's disease and its frequently used abbreviation AD should be represented by a unique NE identifier.

Relation extraction. Relation extraction involves the task of identifying and categorizing semantic relations between named entities in a text. For instance, consider the sentence *Albert Einstein was born in Germany*. The process of relation extraction would need to identify the relation type *born in* between the named entities *Albert Einstein* and *Germany*. The complexity of relation extraction stems from the need not only to identify the entities, but also to understand the context in which they are mentioned to determine their relationship. Machine learning, particularly deep learning approaches such as recurrent neural networks (RNNs) and transformers, have helped achieve top results in this area (Goldberg & Hirst, 2017). These methods often rely on large annotated corpora to train models that can generalize and extract relationships from unseen data.

Document summarization. Document summarization is an information retrieval task that aims to create a concise and coherent summary while preserving the details of the original document. There are two types of document summarization: extractive and abstractive text summaries. Extractive text summarization selects sentences or phrases directly from the source document to create the summary. In contrast, abstractive text summarization goes a step further by interpreting the content and generating new sentences that capture the main ideas. With the advent of deep learning and transformer architectures, abstractive text summarization techniques have greatly improved (Xu, 2022). These techniques add value by allowing users to quickly grasp the essence of a document without having to read it in its entirety. This proves particularly beneficial in areas such as news, research, and business reports.

3.2 Natural language processing

Some key text analysis concepts include the notions of *corpus*, *document*, and *features*. A corpus, also referred to as *document collection*, is a set of text documents from which patterns can be extracted. Corpora can vary greatly in size, from a few hundred to millions, where a corpus of a larger size is more likely to yield more solid patterns.

In the data preparation phase, which is referred to as *text preprocessing*, the goal is to take a set of documents and convert them into a table of feature vectors. There are several methods to embed text into numerical vectors, which are presented in Section 3.5. Historically, in the vector space model (Salton et al., 1975), each feature represents an attribute of a document. In the preprocessing step, each document is represented by a set of features, converted into a sparse vector representation, where each row in the constructed data table represents one document.

If a document is unstructured, its features might be derived through preprocessing methods alone, although some may also be assigned to it, such as author, keywords, or date of publication. The individual features of a document are *words*, *terms*, or *concepts* that occur in a given document. *Words*, which are extracted and tokenized from the text, are the simplest features, with the least amount of semantics. *Terms* are sets of one or more words that occur together, after being normalized, filtered, and grouped, through the use of various term extraction methods. *Concepts* are the most semantics-loaded kind of features: they include terms, but also related notions not necessarily mentioned in a document (for example, a document describing an object, such as a computer, might not make use of the word *computer*; however, the concept *computer* can be assigned to such a document as a feature). The features based on words are usually *sparse* (Feldman & Sanger, 2006). This is because each document contains only relatively few words of the entire document corpus.

Early approaches to text analysis implemented long (and often fragile) pipelines of several different natural language processing (NLP) tasks.

Tokenization. Tokenization splits text into small meaningful units such as words.

Stop-word removal. Stop-word removal removes stop-words that are deemed irrelevant for the given task, e.g., prepositions, which are part of every document, are irrelevant for finding similar documents.

Lemmatization and stemming. Lemmatization is the process of converting one or several inflected forms of a word into a single, 'morphologically neutral' unit, called *lemma*. This process enables different forms of a word to be considered semantically identical, at least for the needs of a given task. For example, the words *discover*, *discovering* and *discovered* all correspond to the lemma *discover*. *Stemming* is a specific approach to converting works to their neutral form, such that a word is truncated to its stem.

Part-of-Speech (POS) tagging. POS tagging is the process of annotating words according to their Part-of-Speech category (e.g., noun, verb), based on the context in which they occur.

Dependency parsing. Dependency parsing extracts a dependency parse of a sentence that represents its grammatical structure and defines the relationships between 'head' words and words, which modify these heads.

Syntactic parsing. Syntactic parsing analyzes sentences syntactically according to a specific grammar.

BoW vector construction. The standard vector representation of text units is the Bag of Words (BoW) vector representation, which transforms the given text into a table of sparse numeric vectors, presented in more detail in Section 3.3.1.

Traditionally, high-quality features that are informative for a given task were difficult to acquire. Classical approaches countered this by incorporating background knowledge about the documents and their domain. This can be achieved by using a controlled vocabulary composed of all terms deemed relevant for a given domain. In many scientific domains, apart from a controlled vocabulary, pre-annotation of large amounts of scientific articles was also used to facilitate the creation of feature vectors.

3.3 Text mining

Text mining (Feldman & Sanger, 2006) is a research area that deals with the construction of models and patterns from text resources, aimed at solving tasks such as text categorization and clustering, taxonomy construction, and sentiment analysis. While this subarea of text analytics is usually considered as a subfield of data mining (DM) research (Han & Kamber, 2001), it could be more generally seen as a multidisciplinary field drawing its techniques from data mining, machine learning, natural language processing (NLP), information retrieval (IR), information extraction (IE) and knowledge management. The standard Bag of Words vector representation of texts is presented in Section 3.3.1, followed by Section 3.3.2 presenting the basic similarity measures used in typical text mining tasks, including document clustering.

3.3.1 Bag of Words vector representations for text units

For text mining, text data needs to be structured as a table of instances and attributes, which can later be used in the modeling phase. The standard vector representation of text units is referred to as the Bag of Words (BoW) vector representation (Sebastiani, 2002). BoW representations of given text units (words, sentences, paragraphs, or documents) result in *sparse* numeric vector representations. Note that alternatively, automated approaches to construct *dense* vector representations can be used to represent texts. Dense vector representations of text units are usually referred to as *embeddings* (see Section 3.5). Embeddings alleviate the need for manual construction of features and build dense representations of text units. An application of embeddings in the context of LBD is illustrated in Chapter 9.

In different text analysis tasks, the individual features (words or terms) in the n-dimensional Bag of Words (BoW) vector representation shall not be equally important. To address this issue, weighting schemes assign importance to features. The four most popular weighting schemes are listed below.

Binary. In this simplest weighting model, a feature weight is 1 if the corresponding feature (word or term) is present in the document; otherwise, it is zero.

Term occurrence. In this model, a feature weight is equal to the number of its occurrences. This weight is often better than a simple binary value since frequently occurring features are more relevant to the given text.

Term Frequency (TF). In this model, the weight is derived from the term occurrence by dividing the vector components by the sum of all vector's weights. The reasoning is similar to the term occurrence, with normalization with respect to the document size.

Term Frequency-Inverse Document Frequency (TF-IDF). In this scheme, the weight of a feature (a word or a term) does not depend only on the document but also on the corpus to which the document belongs. The term weight increases with the number of term occurrences in the document (TF), as this indicates its relevance for the given document. This effect is counterbalanced by the inverse of its frequency (IDF) in the document corpus, since a high frequency in all corpus documents indicates that the term commonly appears in the corpus and not necessarily the term's relevance in a document (Jones, 1972; Salton & Buckley, 1988). TF-IDF is defined as:

$$\text{TF-IDF}(t, d) = \text{TF}(t, d) \cdot \log \frac{|D|}{\text{DF}(t)}, \tag{3.1}$$

where $\text{TF}(t, d)$ is the frequency of the given term t inside document d, $|D|$ is the number of all documents in document corpus D, and $\text{DF}(t)$ is the number of documents that contain term t. The reasoning behind the TF-IDF weighting is to decrease the weights of words/terms that appear in many documents, as this is usually an indication of them being less important.

3.3.2 Similarity measures for text mining

Using vector representations of text units, one can compute the *similarity* of two units, x and y, using some vector similarity measure. Once document similarity has been computed, documents can be grouped into clusters based on their similarity. Note that *similarity measures* and *distance measures* can be used interchangeably, as maximizing the similarity is equivalent to minimizing the distance, and vice versa.

Similarity and distances. While there are numerous measures available for computing vector similarity (Aggarwal, 2015), the standard measure to compute the similarity of BoW vectors used in text analytics is the *cosine similarity* measure, which computes the cosine of angle α between the two non-zero vectors. Its computation involves the computation of the inner product of the two vectors normalized to both have length 1. The *cosine similarity* is defined as follows:

$$\text{cosine}(x, y) = \frac{x \cdot y}{|x| \cdot |y|} = \frac{\sum_{i=1}^{n} x_i \cdot y_i}{\sqrt{\sum_{i=1}^{n} x_i^2} \cdot \sqrt{\sum_{i=1}^{n} y_i^2}}, \tag{3.2}$$

where x_i and y_i are components of vectors x and y, respectively. The cosine similarity measure can be used to find similar documents or documents similar to a given query, provided that the query is also represented as a vector.

There are numerous other measures available to measure the similarity or dissimilarity (distances) of n-dimensional vectors (Aggarwal, 2015), including the

Euclidean and Manhattan distance measures, which are the simplest and the most popular. The Euclidean distance is defined as follows:

$$\text{euclidean}(x, y) = \sqrt{\sum_{i=1}^{n} (x_i - y_i)^2}, \tag{3.3}$$

while the Manhattan distance is defined as follows:

$$\text{manhattan}(x, y) = \sum_{i=1}^{n} |x_i - y_i|. \tag{3.4}$$

Document clustering. A particular clustering tool using TF-IDF document representation and cosine similarity heuristic, named OntoGen (Fortuna et al., 2006), was used in several experiments presented in this monograph. OntoGen is essentially a text mining tool for grouping documents into clusters. The underlying methodology is k-means clustering, a particularly popular technique since only the k parameter needs to be chosen to determine the number of categories into which documents should be clustered. A particularly attractive feature is OntoGen's user-friendly human-computer interface, which can be viewed as a topic detection tool, aimed at automatically creating a topic ontology of topics (concepts). The 'main window' provides a visualization of the ontology, where each concept is represented by the first three keywords (automatically assigned names of clusters that can be edited manually), while the 'concept hierarchy' window provides a quick overview of all concepts with their position in the concept hierarchy, which can also be manipulated directly. While usually used for document clustering and topic identification, the use of OntoGen for outlier document detection is described in Section 6.3 and illustrated in Figure 6.4.

3.4 Network analysis

The field of *network analysis* has its roots in two research fields: mathematical graph theory and social sciences. Network analysis started as an independent research discipline in the late seventies (Zachary, 1977) and early eighties (Burt & Minor, 1983), when sociologists became increasingly aware that the study of social relations—and not only individual attributes—is necessary for in-depth analysis of human societies. Network analysis uses concepts from graph theory to investigate characteristics of networked structures in terms of nodes (such as individual actors, people, or any other objects in the network) and edges/links that connect them (relationships, interactions, or ties between the objects). Since early network analysis research, which focused on *social networks*, where nodes represent people and connections represent their social links, network analysis has grown substantially. The field now covers not only social networks but also other networks originating from any (scientific) discipline. For example, biological networks include gene regulatory networks, protein interaction networks, drug interaction networks, etc. For a comprehensive overview of the field of complex networks, see the reviews by Newman (2003) and Boccaletti et al. (2006).

3.4.1 Network representations

Compared to a standard vector (tabular) data representation, network representations are more complex. Such representations are needed when instances under examination are interconnected to a varying (non-fixed) number of other instances. Representing each connection from an instance as a separate column would result in a different number of columns for each row. Alternatively, if we encoded connections of an instance with a single column, columns would have to contain composite data structures of varying lengths, such as lists.

In mathematical terms, network structures are represented by *graphs*, its nodes are referred to as vertices, and their connections/relations are referred to as edges. Formally, a *graph* $G = (V, E)$ is a mathematical object, composed of a set of vertices (nodes) V, and a set of edges (links) E connecting the vertices. The edges between pairs of vertices (u, v) can be either *directed* or *undirected*, where *undirected graphs* consisting of only undirected edges are considered as a subclass of the class of *directed graphs*.

A graph with no loops (edges connecting a node to itself) and no multiple edges (meaning that a pair of nodes is connected by at most one edge) is called a *simple* graph. *Homogeneous graphs* refer to graphs, where all the nodes and edges are of the same type, while *heterogeneous graphs* refer to graphs composed of various types of nodes and edges.

A way to describe a data set containing inter-connected instances is to represent it as a *network* (Burt & Minor, 1983), a data structure containing nodes and connections/relations between the nodes. Of particular interest are the so-called *information networks*, where the directed connections encode the information flow between the network nodes.

Compared to simpler *homogeneous networks* (including *homogeneous information networks*), more complex types are *heterogeneous information networks* that consist of various types of nodes and links, which became a very popular field of study by Sun and Han (2012). Special heterogeneous directed acyclic graphs that are commonly used for encoding background knowledge are ontologies and knowledge graphs. In information science, an *ontology* (Gruber, 1993; Smith, 2003) is defined as

> an abstract (simplified) view of some selected part of the world, containing the objects, concepts, and other entities that are presumed of interest for some particular purpose and the relationships between them.

Ontologies can be viewed as heterogeneous information networks. As an example, take the Gene Ontology (GO), a frequently used LBD resource, briefly described in Section 2.4.2. Similarly, in *knowledge graphs* (Peng et al., 2003), which are a simpler form of heterogeneous networks, the edges correspond to relations between entities (nodes), and the graph consists of subject-predicate-object *triplets* of the form (h, r, t), where h is referred to as the *head* of the triplet, t as the *tail*, and r as the *relation* connecting the head and the tail.

3.4.2 Network analysis tasks

Graph mining and network analysis cover a wide variety of tasks. Selected network analysis tasks for homogeneous network analysis are outlined below, while heterogeneous network analysis tasks are out of the scope of this overview.

Network node classification. Classification of network data is a natural generalization of classification tasks encountered in a typical machine learning setting. The problem formulation is simple: given a network and class labels for some of the entities in the network, predict the class labels for the rest of the entities in the network. Another name for this problem is *label propagation*. Network node classification in graph-structured data utilizes both labeled and unlabeled data within the network. It considers not only the attributes of individual nodes but also takes into account how they are connected to each other. Label Propagation Algorithm (LPA) is a commonly used algorithm for classifying network nodes. The fundamental concept behind LPA involves propagating labels throughout the network and updating a node label based on the majority of labels among its neighboring nodes. The algorithm works iteratively with each node adopting the label that's most prevalent among its neighbors. This process continues until labels converge (i.e. no further changes occur between iterations) or until a predetermined number of iterations is reached. For example, LPA can be applied in discovering genes associated with a disease. Suppose that researchers have a network where genes are represented as nodes and edges represent known interactions between those genes, such, as protein-protein interactions. Some of these genes have already been identified as being linked to a particular disease (e.g., breast cancer). The objective is to predict the association of the unlabeled genes with the disease (Vanunu et al., 2010).

Link prediction. While classification tasks try to discover new knowledge about network entities, link prediction focuses on unknown connections between the entities. The assumption is that not all network edges are known. The task of link prediction is to predict new edges that are missing or likely to appear in the future. A common approach to link prediction is assigning a score $s(u, v)$ to each pair of vertices u and v which models a probability of the vertices being connected. Link prediction approaches include calculating the score as a product of vertex degrees (Barabási et al., 2002; Newman, 2001), or using the number of common neighbors of two vertices $|N_u \cap N_v|$ (Newman, 2001).

Community detection. While there is no strict definition of the term *network community*, the idea is well summarized by Yang et al., 2010 as follows: a community is a group of network nodes, with dense links within the group and sparse links between the rest of the groups. An extensive overview of community detection methods is provided by Plantié and Crampes, 2013.

Network node ranking. The objective of ranking in information networks is to assess the relevance of a given object either globally (with regard to the whole graph) or locally (relative to some object in the graph). A well-known ranking method is PageRank (Page et al., 1999), which was initially used in the Google search engine. The idea of PageRank (PR) can be explained in two ways.

Random walker. A random walker starts walking from a random vertex v of the network and in each step walks to one of the neighboring vertices with a probability proportional to the weight of the edge traversed. The PageRank of a vertex is then the expected proportion of time the walker spends in the vertex, or, equivalently, the probability that the walker is in the particular vertex after a long time.

Score propagation. The PageRank of a vertex is its score, which it passes to the neighboring vertices. A vertex v_i with a score $PR(i)$ transfers its score to all its neighbors. Each neighbor receives a share of the score proportional to the strength of the edge between it and v_i. This view explains the PageRank with a principle that in order for a vertex to be highly ranked, it must be pointed to by many highly ranked vertices.

Other ranking methods include Personalized PageRank (PPR) (Page et al., 1999), frequently abbreviated as P-PR, which calculates the vertex score locally to a given network vertex, SimRank (Jeh & Widom, 2002), diffusion kernels (Kondor & Lafferty, 2002), hubs and authorities (Chakrabarti et al., 1998) and spreading activation (Crestani, 1997).

Network representation learning. Network representation learning is the process of transforming a given network, its topology, as well as node and edge labels, into a vector representation format (Cui et al., 2018) that can be used in various downstream learning tasks. In the transformed vector representation space, most commonly, rows correspond to individual network nodes, and the columns correspond to automatically constructed features used for describing the properties of nodes in the transformed representation space. More rarely, rows may correspond to edges, and columns to their properties. This task is usually referred to as *network embedding*, briefly mentioned in Section 3.5 below.

3.5 Embeddings

Recent advances in *deep neural networks* (DNNs) (Goodfellow et al., 2016) have led to major breakthroughs in analyzing text, images, speech, and graphs. They are based on the technology of artificial neural networks (NN), in which the information is represented by activation patterns in interconnected networks of primitive units (neurons). This enables concepts to be gradually learned by an NN from the observed data by modifying the connection weights between the hierarchically organized units. Deep neural networks are characterized by several layers of neurons and by training from large data collections. Their successes come at a cost of large storage and computational costs for fast parallel computation, and the need of tuning large amounts of network parameters to avoid data overfitting. The most frequent types of deep neural networks are Long Short Term Memory neural networks (LSTMs) (Hochreiter & Schmidhuber, 1997), convolutional neural networks (CNNs) (LeCun et al., 1998), and transformers (Vaswani et al., 2017). While most common neural network architectures are composed of LSTM or CNN layers at the beginning of a network, followed by one or more fully connected layers, more recent transformer networks use an encoder-decoder architecture, and the attention mechanism, which is highly applicable to text classification and time series analysis.

In addition to DNN, the field of *representation learning* and in particular the embeddings technology have gained popularity (Lavrač et al., 2021). The embedding technology is a prominent side effect of the recent advances in neural networks (NN), where weights that can be extracted from neural networks can be used as a spatial representation that transforms relations between observed entities (data instances) into distances. Embeddings were designed to gain from both worlds: achieving the computational efficiency of standard machine learning and the effectiveness of deep learning. Representation learning through embeddings refers to data transformations of input data into low-dimensional numeric feature vector representations, resulting in decreased number of network neurons, reduced data overfitting, reduced memory requirements, and increased efficiency of learning.

Formally, an embedding is a data transformation, which takes input data of a given data type and format, to construct a tabular representation of the data, where each row represents a single data instance, and each column represents one of the dimensions in the d-dimensional numeric vector space $\mathbb{R}^d$.

Recently, the embedding approach became a prevalent way to build representations of different entity types, e.g., texts, graphs, electronic health records, images, relations, recommendations, etc. We briefly describe some of these data types below.

Text embeddings extract vector representations for words, sentences, and documents from document corpora. The first neural word embeddings like word2vec (Mikolov et al., 2013) produced one vector for each word, regardless of its polysemy (e.g., for a polysemous word like *bank*, word2vec produces a single representation vector and ignores the issue that some words can have different meanings, e.g., a bank is either a financial institution or a land sloping down to a water mass). More recent developments, such as ELMo (Peters et al., 2018) and BERT (Devlin et al., 2019), taking into account the context of a sentence, produce different word vectors for different contexts of each word. Further improvements of neural word embeddings for texts use multitask prediction (inclusion of several related textual prediction tasks).

Network embeddings capture the structure of the graph by using convolution filters which capture the most relevant shapes to predict links in the graph (Bordes et al., 2011; Cui et al., 2018). This way, the weights of the network are used as numerical vectors.

Entity embeddings can use any similarity function between entities to form a prediction task for a neural network. Pairs of entities are used as a training set for the neural network which forecasts whether two entities are similar or dissimilar. The weights of the trained network are then used as its embedding.

In summary, embedding technology is an important side-effect of the recent revival of neural networks (NNs), in which information is represented by activation patterns in interconnected networks of primitive units (neurons). This allows neural networks to gradually learn concepts from observed data by changing the connection weights between hierarchically organized units. These weights, which can be extracted from neural networks, can be used as a spatial representation that converts relationships between observed data entities into distances.

3.6 Large Language Models

Large Language Models (LLMs) can be divided into two groups: Masked Language Models (MLMs) such as BERT (Devlin et al., 2019), and generative autoregressive Language Models such as LLaMa2 (Touvron et al., 2023). These foundational models have established new standards for understanding and generating text with human-level precision (Min et al., 2023).

Building on this ground, sentence transformers (Reimers & Gurevych, 2019) have emerged as a specialized evolution. These models, tailored for the task of learning sentence representations, ensure that semantically similar sentences are closely aligned in the vector space. One notable application of sentence transformers is BERTopic (Grootendorst, 2022), which has revolutionized topic modeling with its unique approach. BERTopic clusters sentences based on semantic similarity, providing a refined and context-sensitive thematic analysis that outperforms conventional methods. In a similar vein, KeyBERT (Grootendorst, 2020) advances the field of keyword extraction. By adopting sentence-transformer technologies, it effectively extracts key terms and phrases from extensive texts (Koloski et al., 2022; Škrlj et al., 2022).

A pivotal area of research in the use of these models is domain adaptation. Wang et al. (2021) proposed an approach for unsupervised domain adaptation, employing sequential denoising auto-encoders to learn from corrupted data. Another approach to domain adaptation involves generative pseudo-labeling (GPL) (Wang et al., 2022), where researchers use a surrogate generative model, such as T5 trained to generate queries for specific passages (Thakur et al., 2021). These queries are then ranked by a cross-encoder (Reimers & Gurevych, 2020) and used as downstream fine-tuning data for the sentence transformer.

The development of prompting techniques in LLMs (Wei et al., 2022), particularly the in-context learning capability of Large Language Models (LLMs) that can perform new tasks by adapting to one (or a few) in-context example(s) (Lampinen et al., 2022), represents a significant stride in model interaction. This approach involves crafting specific prompts that enable models to learn e.g., from a single example within the prompt context, thereby generating more relevant and contextually nuanced responses. This technique, referred to as *one-shot learning*, is crucial in eliciting accurate and specific outputs from models like LLaMa2 (Touvron et al., 2023), demonstrating an advanced level of understanding and flexibility in language generation (Pan et al., 2023).

3.7 Evaluation of LBD approaches

Proper evaluation is crucial to determining the efficacy of newly developed LBD approaches. As elsewhere in science, a proper evaluation method must meet the required criteria, such as: (i) objectivity, (ii) reliability, and (iii) validity.

Cheerkoot-Jalim and Khedo (2021) performed a thorough evaluation of the literature on LBD, employing the guidelines for conducting systematic literature surveys in software engineering (Kitchenham et al., 2016). The synthesis of this work reveals a diverse range of evaluation methods for LBD systems. However, the design of proper evaluation procedures for new LBD methods is still an open issue.

Evaluation of LBD methods has proven to be challenging due to numerous factors, including the difficulty of objectively defining what is a new discovery, and the difficulty of designing numeric measures for assessing how useful, interesting, or actionable is a discovery. There is no gold standard that could be used to accurately benchmark the performance of different LBD approaches, while well-known evaluation metrics (such as precision and recall) cannot be applied without a reference standard. In addition, especially given that novel LBD approaches (especially those relying on embeddings and deep learning) are often very resource-intensive, their resource intensity should be assessed as well. For these reasons, the majority of researchers rely on manual assessment of their discoveries by domain experts to validate the outputs of LBD.

Despite these difficulties, several approaches can be used for LBD systems evaluation. The two most commonly used evaluation approaches are:

Replication of previous discoveries. The most common approach to LBD systems evaluation is through benchmarking of new methods by assessing their LBD potential (e.g., their bridging term discovery or link discovery potential) on standard datasets, used in previous LBD research; we refer the reader to Section 4.1 to get access to standard LBD datasets that can be used to test the capacity of systems for LBD results replication.

Time slicing. Another approach is to apply a simplified cross-validation procedure, by time-slicing the available data set into a training set used for LBD discovery, and a separate test set used for the assessment of discovered artifacts.

Another important issue is the accuracy of the resources and tools embedded in the LBD system. For example, in LBD approaches based on text mining, we rely on entity recognition and relation extraction tools, however, these are inherently error-prone due to inaccuracies in natural language processing. For example, SemRep, a standard NLP tool that extracts semantic relations from PubMed titles and abstracts, yields only 0.55 precision with 0.34 recall (Kilicoglu et al., 2020).

Let us present selected measures that can be used to evaluate the success of the ingredients of LBD methods when applied in the classification and link prediction settings. The basic measures to assess the classification performance of tools embedded in the LBD system are outlined Section 3.7.1, followed by the ROC space evaluation model (Section 3.7.2), the evaluation measures used in link prediction (Section 3.7.3), and the assessment of LBD tools as creativity support tools (Section 3.7.4).

3.7.1 Evaluation measures used in classification

In machine learning, the evaluation of machine-learned classifiers is performed on a test set that was not used to train the model. For small data sets, where a hold-out set would significantly hinder the learning capability due to wasted training data, the cross-validation approach to classifier evaluation is performed.

In a binary classification problem, let E denote the set of all training instances, P denote the set of positive instances, and N the set of negative instances, where $P \cup N = E$ and $|P| + |N| = |E|$. Let $TP \subseteq P$ (true positives) be a set of positive instances that are correctly classified by the learned model, $TN \subseteq N$ (true negatives) be a set of correctly classified negative instances, $FP \subseteq N$ (false positives) be a set of negative instances that are incorrectly classified as positives by the learned model, and $FN \subseteq P$ (false negatives) be a set of positive instances incorrectly classified as negative instances. These notions are explained in Table 3.1.

Typical metrics used in classification are classification accuracy, classification error, precision, recall, and the F-measure defined below.

Classification accuracy. The classification quality of the learned models is measured by the classification accuracy, which is defined as the proportion of the total number of correctly classified examples in all classes relative to the total number of tested examples. As a special case, for a binary classification problem, let $TP \subseteq P$ be a set of positive examples that are correctly classified by the learned model, and $TN \subseteq N$ be a set of correctly classified negative examples. In this case, the accuracy of a model is computed as follows:

$$\text{Accuracy} = \frac{|TP| + |TN|}{|E|},$$

where $|TP|$ and $|TN|$ denote the number of correctly classified positive and negative examples. Note that Accuracy estimates the accuracy of the classifier on both positive and negative examples of the target class of interest. Its range is $[0, 1]$, where higher values denote a better accuracy. Instead of accuracy, results are often presented with an equivalent *classification error* measure, which is simply Error = 1 − Accuracy.

Precision. In binary class problems, precision is the ratio between the number of correctly classified positive instances and all predicted as positives, i.e.

$$\text{Precision} = \frac{|TP|}{|TP| + |FP|}. \tag{3.5}$$

Table 3.1 Confusion matrix depicting the notation for sets of actual and predicted positive and negative examples.

Actual	Predicted		All
	Positive	Negative	
Positive examples	# True Positives $\|TP\|$	# False Negatives $\|FN\|$	$\|P\|$
Negative examples	# False Positives $\|FP\|$	# True Negatives $\|TN\|$	$\|N\|$

Recall. Recall (also known as sensitivity) in binary problems is the ratio of positive instances to the number of all positive instances.

$$\text{Recall} = \frac{|TP|}{|TP| + |FN|} = \frac{|TP|}{|P|}. \tag{3.6}$$

F-measure. A measure that combines precision and recall in binary problems as a harmonic mean of precision and recall is called the F_1 measure or the balanced F-score. It is defined as follows.

$$F_1 = \frac{2 \cdot \text{Precision} \cdot \text{Recall}}{\text{Precision} + \text{Recall}}. \tag{3.7}$$

The range of precision, recall, and F_1 is $[0, 1]$, and larger values denote better performance.

3.7.2 The ROC space model

ROC analysis[1] has its origins in signal detection theory. A ROC curve is a two-dimensional plot for analyzing the operating characteristics of the receiving station in terms of its hit and false alarm rates. The original ROC space model has been adapted in a straightforward way to classification as a framework to select the best classifier, where a point in the ROC space shows the performance of a classifier in terms of *false positive rate* (FPR) (plotted on the x-axis), and sensitivity or *true positive rate* (TPR) (plotted on the y-axis). In binary classification, a perfect classifier that makes no mistakes is plotted in the upper left corner, at point $(0, 1)$, correctly classifying all positive and no negative examples. The origin $(0, 0)$ represents a useless classifier that classifies no examples as positive, whereas the point $(1, 1)$ corresponds to the universal classifier, which classifies all examples as positive, respectively. If all classifiers are plotted in this ROC space, the best rule is the one closest to $(0, 1)$, where the distance measure used depends on the relative misclassification costs of the two classes.

ROC analysis can be used to evaluate both classification and ranking algorithms. For a more detailed account of the ROC analysis for classification and ranking, we refer to Fawcett (2006).

Evaluating classifiers. The ROC space is appropriate to evaluate the quality of binary classifiers, as the classifiers plotted near the diagonal can be discarded as insignificant. The reason is that the points on the diagonal represent classifiers whose true positive rate equals their false positive rate: their success rate on the positive examples thus equals their error rate on the negative examples, and vice versa. The classifier's distance from the diagonal can thus be used as a quality measure. The farther a classifier is away from the diagonal, the better its ability to discriminate between positive and negative examples. The perfect classifier in the upper left corner of the ROC space has a maximum distance from the diagonal.

[1] ROC stands for *Receiver Operating Characteristic*.

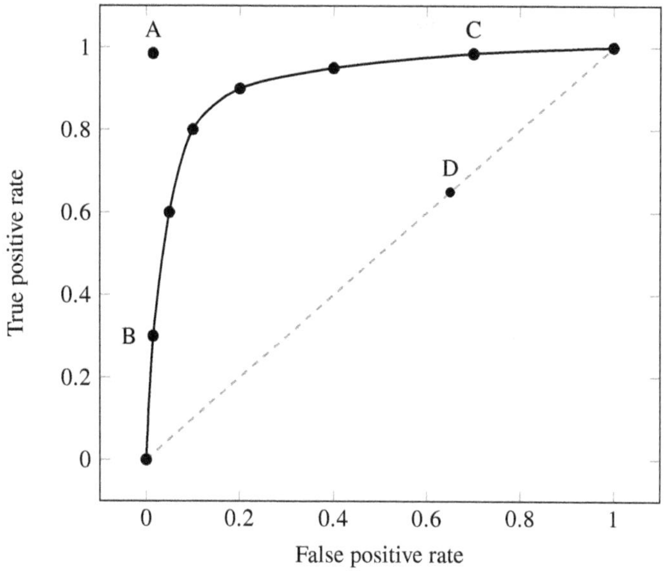

Fig. 3.1 The ROC space model with classifier performances: (A) nearly perfect, (B) conservative, (C) liberal, and (D) random guess classifier, adapted from Fawcett (2006).

One can also show that all rules that are optimal under some (linear) cost model must lie on the *convex hull* of a set of points in the ROC space (Fawcett, 2006). Essentially, the convex hull starts at the point $(0,0)$ and ends at $(1,1)$, and encloses all the points so that no straight line connecting two points will be outside the convex hull. This hull is also known as the *ROC curve*. The ROC space and the convex hull are illustrated in Figure 3.1. A classifier that lies below this convex hull can never be optimal under any linear cost model and can thus be removed from further consideration.

Evaluating ranking algorithms. ROC curves can also be used to evaluate ranking algorithms. The key idea behind this is to draw the ROC curve, where each point in the curve corresponds to a threshold in the predicted ranking. The ideal ranker would rank all positive examples before all negative examples, while the worst ranker would rank the instances randomly, which corresponds to a ROC curve that moves along the diagonal. To evaluate the quality of a ranker, one can compute the *area under the ROC curve (AUC)*, which would be 1 for the ideal ranker and $\frac{1}{2}$ for a random ranker. The AUC corresponds to the probability that a randomly selected positive example is ranked before a randomly selected negative example.

Note that AUC analysis can also be adapted to evaluate the ingredients of classification or ranking algorithms, including the evaluation of heuristics used as ingredients of LBD systems. For example, the heuristics used for *b*-term detection in the CrossBee LBD system included over 20 elementary heuristics and numerous combined (ensemble-based) heuristics used to assess and rank the detected *b*-terms (details on the use of ROC curves to evaluate the quality of CrossBee heuristics are presented in Section 5.2).

3.7.3 Evaluation measures used in link prediction

In link prediction problems, the task is to predict an entity given a relation and another entity, i.e. we want to infer h given $(?, r, t)$ or infer t given $(h, r, ?)$ where ? refers to a missing entity. Knowledge graph completion models are usually automatically assessed using an evaluation protocol proposed by Bordes et al. (2013). Suppose that $\mathcal{X}$ is a set of triples, Θ_E the embeddings of entities $\mathcal{E}$, and Θ_R the embeddings of relations $\mathcal{R}$. In the first corruption step, we go through a set of triples and for each triple $\mathbf{x} = (h, r, t) \in \mathcal{X}$ we replace its head and tail with all other entities in $\mathcal{E}$. Each triple is corrupted exactly $2|\mathcal{E}| - 1$ times. Formally, the corrupted triple is defined as:

$$\widetilde{\mathbf{x}} = \bigcup_{h' \in \mathcal{E}} (h', r, t) \cup \bigcup_{t' \in \mathcal{E}} (h, r, t'),$$

where $h' \neq h$ and $t' \neq t$. We employ the filtered setting protocol that does not take into account any corrupted triple that already appears in a knowledge graph $\mathcal{G}$. In the second scoring phase, original and corrupted triples are tested using the constructed classifier ψ. The intuition behind this is that the model will assign a higher score to the original triple and a lower score to the corrupted triple. In the third evaluation phase, the proposed models are assessed using three measures: mean rank (MR), mean reciprocal rank (MRR), and Hits@k.

Mean rank (MR). MR is an average rank assigned to the true relation over all the relations in a test set $\mathcal{T}$:

$$\mathrm{MR} = \frac{1}{2|\mathcal{T}|} \sum_{i=1}^{|\mathcal{T}|} \left(\mathsf{rank}_i^h + \mathsf{rank}_i^t\right), \tag{3.8}$$

where rank_i^h and rank_i^t denote the rank position:

$$\mathsf{rank}_i^h = 1 + \sum_{\widetilde{\mathbf{x}}_i \in C^h(\mathbf{x}_i) \setminus \mathcal{G}} I\left[\psi(\mathbf{x}_i) < \psi(\widetilde{\mathbf{x}}_i)\right]$$

$$\mathsf{rank}_i^t = 1 + \sum_{\widetilde{\mathbf{x}}_i \in C^h(\mathbf{x}_i) \setminus \mathcal{G}} I\left[\psi(\mathbf{x}_i) < \psi(\widetilde{\mathbf{x}}_i)\right],$$

where C is the set of corrupted triples and the indicator function $I[P]$ is 1 iff P is true, and 0 otherwise.

Mean reciprocal rank (MRR). MRR is the average inverse rank for all test triples and is formally computed as:

$$\mathrm{MRR} = \frac{1}{2|\mathcal{T}|} \sum_{\mathbf{x}_i \in \mathcal{T}} \frac{1}{\mathsf{rank}_i^h} + \frac{1}{\mathsf{rank}_i^t}. \tag{3.9}$$

Hits@k. Hits@k measures the percentage of relations in which the true triple appears in the top k ranked triples, where $k \in \{1, 3, 10\}$; formally:

$$\mathrm{Hits@}k = \frac{100}{2|\mathcal{T}|} \sum_{\mathbf{x}_i \in \mathcal{T}} I\left[\mathsf{rank}_i^h \leq k\right] + I\left[\mathsf{rank}_i^t \leq k\right]. \tag{3.10}$$

The goal is to achieve low MR, and high MRR and Hits@k.

3.7.4 Qualitative evaluation of LBD support systems

LBD is typically not an automated process and assumes human-in-the-loop in the process of using online LBD tools, where Human-Computer Interaction (HCI) interfaces are crucial for successful discovery. Hence, from the HCI perspective, LBD tools can also be viewed as *creativity support tools* (CST), whose design aims to develop improved software and user interfaces that empower users to be both more productive and more innovative. Following the set of design principles for CSTs by Shneiderman (2007, 2009) outlined in Section 1.3.1, the success of an LBD system can also be evaluated using the following qualitative criteria:

- Providing support in exploration, discovery, and innovation, through effectively organizing search results by ranking, clustering, and partitioning with ample tools for annotation, tagging, and marking.
- Enabling collaboration, given that the processes that lead to an individual's serendipitous discoveries are often highly collaborative.
- Providing rich history-keeping by keeping a record of which alternatives they have tried, enabling them access to the earlier alternatives to make modifications.
- Supporting a steep learning curve for novices (low threshold), yet providing sophisticated functionality that experts need (high ceilings), and also delivering a wide range of supplementary services to choose from (wide walls).

References

Aggarwal, C. (2015). *Data Mining: The Textbook*. Springer.

Barabási, A. L., Jeong, H., Néda, Z., Ravasz, E., Schubert, A., & Vicsek, T. (2002). Evolution of the social network of scientific collaborations. *Physica A: Statistical Mechanics and its Applications*, *311*(3-4), 590–614.

Beliga, S., Meštrović, A., & Martinčić-Ipšić, S. (2015). An overview of graph-based keyword extraction methods and approaches. *Journal of Information and Organizational Sciences*, *39*(1), 1–20.

Boccaletti, S., Latora, V., Moreno, Y., Chavez, M., & Hwang, D. (2006). Complex networks: Structure and dynamics. *Physics Reports*, *424*(4), 175–308.

Bordes, A., Usunier, N., Garcia-Duran, A., Weston, J., & Yakhnenko, O. (2013). Translating embeddings for modeling multi-relational data. *Proceedings of the 26th International Conference on Neural Information Processing Systems*, *26*, 2787–2795.

Bordes, A., Weston, J., Collobert, R., & Bengio, Y. (2011). Learning structured embeddings of knowledge bases. *Proceedings of the AAAI Conference on Artificial Intelligence*, *25*(1), 301–306.

Burt, R. S., & Minor, M. J. (1983). *Applied Network Analysis: A Methodological Introduction*. Sage Publications.

Chakrabarti, S., Dom, B., Raghavan, P., Rajagopalan, S., Gibson, D., & Kleinberg, J. (1998). Automatic resource compilation by analyzing hyperlink structure and associated text. *Computer Networks*, *30*(1–7), 65–74.

Cheerkoot-Jalim, S., & Khedo, K. K. (2021). Literature-based discovery approaches for evidence-based healthcare: A systematic review. *Health and Technology*, *11*(6), 1205–1217.

Crestani, F. (1997). Application of spreading activation techniques in information retrieval. *Artificial Intelligence Review*, *11*(6), 453–482.

Cui, P., Wang, X., Pei, J., & Zhu, W. (2018). A survey on network embedding. *IEEE Transactions on Knowledge and Data Engineering*, *31*(5), 833–852.

Devlin, J., Chang, M.-W., Lee, K., & Toutanova, K. (2019, June). BERT: Pre-training of deep bidirectional transformers for language understanding. In J. Burstein, C. Doran, & T. Solorio (Eds.), *Proceedings of the 2019 Conference of the North American Chapter of the Association for Computational Linguistics: Human Language Technologies* (pp. 4171–4186, Vol. 1). Association for Computational Linguistics.

Fawcett, T. E. (2006). An introduction to ROC analysis. *Pattern Recognition Letters*, *27*(8), 861–874.

Feldman, R., & Sanger, J. (2006). *The text mining handbook: Advanced approaches in analyzing unstructured data*. Cambridge University Press.

Fortuna, B., Grobelnik, M., & Mladenić, D. (2006). Semi-automatic data-driven ontology construction system. *Proceedings of the 9th International Multiconference Information Society*, 223–226.

Goldberg, Y., & Hirst, G. (2017). *Neural Network Methods for Natural Language Processing*. Morgan & Claypool.

Goodfellow, I., Bengio, Y., & Courville, A. (2016). *Deep Learning*. The MIT Press.

Grootendorst, M. (2020). KeyBERT: Minimal keyword extraction with BERT.

Grootendorst, M. (2022). BERTopic: Neural topic modeling with a class-based TF-IDF procedure.

Gruber, T. R. (1993). A translation approach to portable ontology specifications. *Knowledge Acquisition*, *5*(2), 199–220.

Han, J., & Kamber, M. (2001). *Data Mining: Concepts and Techniques*. Morgan Kaufmann Publishers.

Hochreiter, S., & Schmidhuber, J. (1997). Long short-term memory. *Neural Computation*, *9*(8), 1735–1780.

Hoffart, J., Yosef, M. A., Bordino, I., Fürstenau, H., Pinkal, M., Spaniol, M., Taneva, B., Thater, S., & Weikum, G. (2011). Robust disambiguation of named entities in text. *Proceedings of the Conference on Empirical Methods in Natural Language Processing*, 782–792.

Jeh, G., & Widom, J. (2002). SimRank: A measure of structural-context similarity. *Proceedings of the 8th ACM SIGKDD International Conference on Knowledge Discovery and Data Mining*, 538–543.

Jones, K. S. (1972). A statistical interpretation of term specificity and its application in retrieval. *Journal of Documentation*, *28*, 11–21.

Kilicoglu, H., Rosemblat, G., Fiszman, M., & Shin, D. (2020). Broad-coverage biomedical relation extraction with SemRep. *BMC Bioinformatics*, *21*(1), 188.

Kitchenham, B. A., Budgen, D., & Brereton, P. (2016). *Evidence-Based Software Engineering and Systematic Reviews*. CRC Press.
OCLC: ocn932588149.

Koloski, B., Pollak, S., Škrlj, B., & Martinc, M. (2022). Out of thin air: Is zero-shot cross-lingual keyword detection better than unsupervised? *Proceedings of the Thirteenth Language Resources and Evaluation Conference*, 400–409.

Kondor, R. I., & Lafferty, J. D. (2002). Diffusion kernels on graphs and other discrete input spaces. *Proceedings of the 19th International Conference on Machine Learning*, 315–322.

Lampinen, A., Dasgupta, I., Chan, S., Mathewson, K., Tessler, M., Creswell, A., McClelland, J., Wang, J., & Hill, F. (2022). Can language models learn from explanations in context? In Y. Goldberg, Z. Kozareva, & Y. Zhang (Eds.), *Findings of the Association for Computational Linguistics: EMNLP 2022* (pp. 537–563). Association for Computational Linguistics.

Lavrač, N., Podpečan, V., & Robnik-Šikonja, M. (2021). *Representation learning: Propositionalization and embeddings*. Springer.

LeCun, Y., Bottou, L., Bengio, Y., & Haffner, P. (1998). Gradient-based learning applied to document recognition. *Proceedings of the IEEE*, *86*(11), 2278–2324.

Manning, C., Raghavan, P., & Schütze, H. (2008). *Introduction to Information Retrieval*. Cambridge University Press.

Mikolov, T., Chen, K., Corrado, G., & Dean, J. (2013). Efficient estimation of word representations in vector space. *arXiv preprint 1301.3781*.

Min, B., Ross, H., Sulem, E., Veyseh, A. P. B., Nguyen, T. H., Sainz, O., Agirre, E., Heintz, I., & Roth, D. (2023). Recent advances in natural language processing via large pre-trained language models: A survey. *ACM Computing Surveys*, *56*(2), 1–40.

Newman, M. E. J. (2001). The structure of scientific collaboration networks. *Proceedings of the National Academy of Sciences of the United States of America*, *98*(2), 404–409.

Newman, M. E. J. (2003). The Structure and Function of Complex Networks. *SIAM Review*, *45*(2), 167–256.

Newman, M. (2001). Clustering and preferential attachment in growing networks. *Physical Review E*, *64*(2), 025102.

Page, L., Brin, S., Motwani, R., & Winograd, T. (1999). *The PageRank citation ranking: Bringing order to the Web.* (tech. rep.). Stanford InfoLab. Stanford InfoLab.

Pan, J., Gao, T., Chen, H., & Chen, D. (2023). What in-context learning "learns" in-context: Disentangling task recognition and task learning. In A. Rogers, J. Boyd-Graber, & N. Okazaki (Eds.), *Findings of the Association for Computational Linguistics: ACL 2023* (pp. 8298–8319). Association for Computational Linguistics.

Peng, C., Xia, F., Naseriparsa, M., & Osborne, F. (2003). Knowledge Graphs: Opportunities and Challenges. *Artificial Intelligence Review*, *56*, 13071–13102.

Peters, M. E., Neumann, M., Iyyer, M., Gardner, M., Clark, C., Lee, K., & Zettlemoyer, L. (2018). Deep contextualized word representations. *arXiv preprint:1802.05365*.

Plantié, M., & Crampes, M. (2013). Survey on social community detection. In N. Ramzan, R. Zwol, J.-S. Lee, K. Clüver, & X.-S. Hua (Eds.), *Social Media Retrieval* (pp. 65–85). Springer London.

Reimers, N., & Gurevych, I. (2019). Sentence-BERT: Sentence embeddings using siamese BERT-networks. *Proceedings of the 2019 Conference on Empirical Methods in Natural Language Processing*. https://arxiv.org/abs/1908.10084

Reimers, N., & Gurevych, I. (2020). Making monolingual sentence embeddings multilingual using knowledge distillation. *Proceedings of the 2020 Conference on Empirical Methods in Natural Language Processing*. https://arxiv.org/abs/2004.09813

Salton, G., & Buckley, C. (1988). Term-weighting approaches in automatic text retrieval. *Information Processing and Management*, *24*(5), 513–523.

Salton, G., Wong, A., & Yang, C. S. (1975). A vector space model for automatic indexing. *Communications of the ACM*, *18*(11), 613–620.

Sebastiani, F. (2002). Machine learning in automated text categorization. *ACM Computing Surveys*, *34*(1), 1–47.

Shneiderman, B. (2007). Creativity support tools: Accelerating discovery and innovation. *Communications of the ACM*, *50*(12), 20–32.

Shneiderman, B. (2009). Creativity support tools: A grand challenge for HCI researchers. In *Engineering the User Interface* (pp. 1–9). Springer.

Škrlj, B., Koloski, B., & Pollak, S. (2022). Retrieval-efficiency trade-off of unsupervised keyword extraction. In P. Pascal & D. Ienco (Eds.), *Proceedings of the 25th International Conference on Discovery Science* (pp. 379–393). Springer Nature Switzerland.

Smith, B. (2003). Chapter 11: Ontology. In L. Floridi (Ed.), *Blackwell guide to the philosophy of computing and information* (pp. 155–166, Vol. 7250). Blackwell.

Sun, Y., & Han, J. (2012). *Mining Heterogeneous Information Networks: Principles and Methodologies*. Morgan & Claypool Publishers.

Thakur, N., Reimers, N., Rücklé, A., Srivastava, A., & Gurevych, I. (2021). BEIR: A heterogeneous benchmark for zero-shot evaluation of information retrieval models. *Proceedings of the 35th Conference on Neural Information Processing Systems Track on Datasets and Benchmarks, 1*.

Touvron, H., Martin, L., Stone, K., Albert, P., Almahairi, A., & Babaei, Y. (2023). *Llama 2: Open Foundation and Fine-Tuned Chat Models*. ArXiv. https://doi.org/10.48550/arXiv.2307.09288

Vanunu, O., Magger, O., Ruppin, E., Shlomi, T., & Sharan, R. (2010). Associating genes and protein complexes with disease via network propagation. *PLoS Computational Biology*, *6*(1).

Vaswani, A., Shazeer, N., Parmar, N., Uszkoreit, J., Jones, L., Gomez, A. N., Kaiser, Ł., & Polosukhin, I. (2017). Attention is all you need. *Advances in Neural Information Processing Systems*, *30*, 5998–6008.

Wang, K., Reimers, N., & Gurevych, I. (2021). TSDAE: Using transformer-based sequential denoising auto-encoderfor unsupervised sentence embedding learning. *Findings of the Association for Computational Linguistics: EMNLP 2021*, 671–688.

Wang, K., Thakur, N., Reimers, N., & Gurevych, I. (2022). GPL: Generative pseudo labeling for unsupervised domain adaptation of dense retrieval. *Proceedings of the 2022 Conference of the North American Chapter of the Association for Computational Linguistics: Human Language Technologies*, 2345–2360.

Wei, J., Wang, X., Schuurmans, D., Bosma, M., Xia, F., Chi, E., Le, Q. V., Zhou, D., et al. (2022). Chain-of-thought prompting elicits reasoning in large language models. *Advances in Neural Information Processing Systems*, *35*, 24824–24837.

Xu, J. (2022). *Neural Text Summarization with Fine-grained Control*. University of Texas.

Yang, B., Liu, D., & Liu, J. (2010). Discovering communities from social networks: Methodologies and applications. In B. Furht (Ed.), *Handbook of Social Network Technologies and Applications* (pp. 331–346). Springer USA.

Zachary, W. W. (1977). An information flow model for conflict and fission in small groups. *Journal of Anthropological Research*, *33*, 452–473.

Chapter 4
Benchmark Data and Reusable Python Code

This chapter presents the open science approach to literature-based discovery implemented in this monograph. Section 4.1 provides access to several datasets that have been used in the original LBD articles to illustrate the potential of early and more recent LBD approaches to replicate the reported discoveries. Section 4.2 provides selected Python components that constitute specific LBD pipelines that can be reused to illustrate the original LBD approaches. These pipelines, implemented in the form of Python Jupyter notebooks, are complemented by more elaborate LBD pipelines developed within the individual chapters of the monograph, explained in *Tutorials in Python* sections, included in most of the subsequent chapters of this monograph.

4.1 Benchmark data sets used for experiment replicability

A selection of datasets prepared for reuse and tool utility verification has been constructed in a way that enables replication of the originally reported LBD experiments. To achieve this, the query to download a PubMed dataset is limited by the year of the paper presenting the originally discovered *bridging terms* (b-terms), which is different for every dataset. The summary of statistics on the datasets used in our experiments is presented in Table 4.1.

Table 4.1 Basic properties and statistics of domain pair datasets made available in this monograph.

Property	DFO-RS	Mig-Mg	AUT-CaN
No. of documents	1273/153	1156/7320	10,819/5320
Mean no. of words/document	13.1/18.3	11.3/16.1	166/259
Mean no. of terms/document	2.56/1.92	11.5/12.1	4.02/4.03
No. of unique terms	4554/1059	1812/6217	423,153/316,702
No. of common terms	27	228	46,473
No. of bridging terms	3	43	13

N. Lavrač et al., *Bisociative Literature-Based Discovery*,
https://doi.org/10.1007/978-3-031-96863-1_4

Table 4.2 Bridging terms (*b*-terms) identified by Swanson (1986, 1988) and Petrič et al. (2009) for the DFO-RS, Mig-Mg, and AUT-CaN domain pairs, respectively.

Dietary Fish Oil-Raynaud's Syndrome (DFO-RS)
blood viscosity, platelet aggregation, vascular reactivity
Migraine-Magnesium (Mig-Mg)
serotonin, spread, spread depression, seizure, calcium antagonist, vasospasm, paroxysmal, stress, prostaglandin, reactivity, spasm, inflammatory, anti inflammatory, 5 hydroxytryptamine, calcium channel, epileptic, platelet aggregation, verapamil, calcium channel blocker, nifedipine, indomethacin, prostaglandin e1, anticonvulsant, arterial spasm, coronary spasm, cerebral vasospasm, convulsion, cortical spread depression, brain serotonin, 5 hydroxytryptamine receptor, epilepsy, antimigraine, 5 ht, epileptiform, platelet function, prostacyclin, hypoxia, diltiazem, convulsive, substance p, calcium blocker, prostaglandin synthesis, anti aggregation
Autism-Calcineurin (AUT-CaN)
synaptic, synaptic plasticity, calmodulin, radiation, working memory, bcl 2, type 1 diabetes, ulcerative colitis, asbestos, deletion syndrome, 22q11 2, maternal hypothyroxinemia, bombesin

We used bridging terms (*b*-terms), which were identified in each domain pair dataset, as the gold standard to evaluate the utility of the LBD approaches tested for cross-context b-term discovery. Table 4.2 presents the *b*-terms of the *dietary fish oil–Raynaud's syndrom* (DFO-RS), *migraine–magnesium* (Mig-Mg), and *autism–calcineurin* (AUT-CaN) domain pair datasets used in our experiments.

Dietary Fish Oil–Raynauld's disease (DFO-RS) benchmark. Swanson (1986) acquired an initial collection of 489 articles and conducted a Dialog® Scisearch using the terms "Raynaud" and "Fish Oil", specifically searching through the titles and abstracts of MEDLINE and Embase (Excepta Medica) citations. Out of the 489 publications, he discovered that just four of them had cross-references that covered both the dietary fish oil (DFO) and Raymanud's syndrome (RS) papers. Out of these four publications, he discovered that only two (Moncada, 1983; Moncada & Vane, 1984) were actually linked to both RS and DFO.

Numerous authors in the LBD community have relied on the DFO-RS hypothesis (i.e. *b*-terms crossing DFO and RS literatures) as a baseline for evaluating novel LBD approaches, by assessing the efficacy of new methods in identifying and ranking Swanson's bridging terms that link dietary fish oil to Raynaud's syndrome. The three most important terms we focus on are *blood viscosity*, *platelet aggregation*, and *vascular reactivity*.

In our experiments, we used the following PubMed queries to build the DFO-RS dataset:

1. `fish oil [TI] AND 1900/01/01:1985/11/30 [PDAT]`
2. `raynaud* [TI] AND 1900/01/01:1985/11/30 [PDAT]`

Note that Swanson formulated the DFORS hypothesis at the end of November 1985; hence, we limited the upper bound of the time window with the PubMed "Publication Date" (PDAT) tag to reflect the body of scientific literature available at that time.

Migraine–Magnesium (Mig-Mg) benchmark. The second benchmark dataset pertains to the discovery of implicit links between migraine (C) and magnesium (A). Swanson (1988) described eleven neglected connections (b-terms) between sets A and C, explaining magnesium deficiency as an important cause of the onset of migraine disease in humans.

In the LBD process, Swanson (1988) managed to find more than 60 pairs of articles connecting the migraine domain with magnesium deficiency via several bridging concepts. In this process, Swanson identified 43 b-terms connecting the two domains of the *migraine-magnesium* domain pair.[1] For our research, we used two PubMed queries to assemble the Mig-Mg dataset:

1. `migraine [TI] AND 1900/01/01:1987/12/31 [PDAT]`
2. `magnesium [TI] AND 1900/01/01:1987/12/31 [PDAT]`

Autism–Calcineurin (AUT-CaN) benchmark. The third dataset, the autism–calcineurin domain pair, was introduced and initially researched by Petrič et al. (2009). Autism belongs to a group of pervasive developmental disorders that are portrayed by an early delay and abnormal development of the cognitive, communication, and social interaction skills of a person. It is a very complex and not yet sufficiently understood domain, where precise causes are still unknown. Like Swanson (1986, 1988), Petrič et al. (2009) also provided b-terms, 13 in total, whose importance in connecting autism to calcineurin (a protein phosphatase) was discussed and confirmed by the domain expert.

The dataset was constructed using the following PubMed query:

1. `autis* [TIAB] AND 1900/01/01:2007/12/31 [PDAT]`
2. `calcineurin [TIAB] AND 1900/01/01:2007/12/31 [PDAT]`

Preprocessing was performed in a standard text mining way, using the following preprocessing steps: (i) text tokenization, (ii) stop-word removal, (iii) word stemming/lemmatization using the LemmaGen lemmatizer for English, (iv) construction of N-grams (terms defined as a concatenation of 1 to N words that appear consecutively in text, with a minimum support frequency), and (v) creation of the standard Bag of Words (BoW) representation of text using term-frequency-inverse-document-frequency (TF-IDF) weighting. In addition, to enhance the standard workflow, terms containing words that served as query terms during document selection were removed from the dataset to ensure unbiased preprocessing.

[1] In our experiments, we found a direct link between migraine and magnesium, which Vosgerau (1973) discussed in their paper but Swanson (1988) did not report. Swanson conducted his research using what is now known as the Medline subset of PubMed. To replicate Swanson's discovery, it is necessary to limit the Entrez query to the Medline subset only.

4.2 Reusable processing components and LBD pipelines

This section provides an overview of the elementary processing components and sample LBD pipelines that form the backbone for the reproducibility of simplified LBD experiments to illustrate the use of the LBD techniques described in the monograph. The notebooks provided are intended to serve as illustrative examples of the LBD principles. The main focus of these materials is on the repeatability of the experiments rather than on the efficient execution of the Python code. This approach emphasizes the understanding and practical application of concepts so that the reader can consistently repeat the results and effectively grasp the underlying principles.

Historically, the development of modern programming languages, programming paradigms, and operating systems enabled the development of computer platforms for text analysis, and of modern integrated text analysis tools. Such platforms offer a high degree of abstraction and allow the user to focus on analyzing the results rather than on the methods used to obtain the results.

Originally, individual algorithms were implemented as complete solutions for specific natural language processing (NLP) problems, while second-generation NLP systems were intended to provide integrated solutions for entire NLP pipelines including text preprocessing, transformation, and text analysis tasks, ideally also with graphical user interfaces. Many of these, such as scikit-learn (Pedregosa et al., 2011), take advantage of operating system-independent languages, such as the Python environment, to develop complete solutions that include methods for text preprocessing and visual representation of results. These solutions also provide interfaces to call each other and thereby enable cross-solution development.

A common practice in modern scientific publishing is to make code and data publicly available, e.g., via GitHub. Today's AI software is further confronted with the challenge of how to adapt such open science solutions to big data and deep learning, which require grid and GPU processing. Hugging Face (Wolf & et al., 2020) exemplifies open-source collaboration within the NLP community by providing widespread access to state-of-the-art models. Similarly, Re:Science (2025) is a community-driven journal that focuses on republishing and reimplementing previously published methods to promote reproducible and accessible science.

Additional means of simplifying access to NLP tools and enabling cross-solution development are modern web-based Jupyter Python Notebooks (Toomey, 2017) that allow the creation and sharing of documents containing live code, equations and visualizations. The open science approach to literature-based discovery implemented in this monograph uses Jupyter Python Notebooks to practically illustrate selected concepts and LBD pipelines, allowing the reader to run the described methods on illustrative examples. The notebooks provide a set of examples that we have carefully chosen to be easy to understand but are non-trivial and can be used as recipes or starting points for developing more complex LBD scenarios. Moreover, following recommendations and guidelines to ensure reproducible science and FAIR benchmarking (Peng, 2011), we prepared dockerized versions of all methods, which provide off-the-shelf and ready-to-use environment for testing and working with the LBD methods surveyed in this monograph.

4.2.1 Jupyter Python notebooks: Brief installation guidelines

To use the notebooks, the JupyterLab application has to be installed and activated. The following steps provide the complete setup of the environment in which the notebooks can be inspected, edited, run, and reused.

1. **Install the Python language interpreter**. It is recommended to use the latest version of the 3.11 branch to ensure maximum support and stability (older 3.7 to 3.10 branches have also been tested and should also work). The official repository[2] offers a large selection of installers of different versions for the supported platforms.
2. **Download the notebook repository of the monograph**. The repository can be downloaded by the version control system git or by downloading the archive of the selected branch. For example, the archive of the master branch is available at https://github.com/lbd-book/code.git.
3. **Create a virtual environment.** A virtual environment provides an isolated development environment for Python packages. It is recommended to create a separate virtual environment for each notebook to minimize possible conflicts between different versions of the required packages. Use the following commands to create and activate a new virtual environment named 'lbd' in the current folder on a Linux-based operating system (including MacOS):

   ```
   python -m venv lbd
   source lbd/bin/activate
   ```

   On a Windows operating system, the corresponding commands are the following:

   ```
   python -m venv lbd
   lbd/Scripts/activate
   ```

4. **Install Python packages**. Python packages used by the notebooks from the repository of this monograph are installed in bulk according to the configuration file, `requirements.txt`, with the following command:

   ```
   pip install -r requirements.txt
   ```

5. **Install JupyterLab**. JupyterLab is a web-based interactive development environment for Jupyter notebooks, code, and data. You will use it to open, run, and modify notebooks from the repository of this monograph. With the virtual environment activated, the following command will install the latest version of `jupyterlab`:

   ```
   pip install jupyterlab
   ```

6. **Run JupyterLab**. The JupyterLab web application can be launched with the following command:

[2] https://www.python.org/downloads/

```
jupyter lab
```

The command will print a local web address where the JupyterLab web application is running. On most platforms, the page will open automatically in your default web browser.

7. **Open a Jupyter notebook.** JupyterLab allows access to files and subfolders in the folder where it was launched. It is therefore recommended to run Jupyter inside the directory containing the monograph's repository of notebooks.

4.2.2 Running Jupyter Notebooks using Docker

Instead of setting up a Python environment manually, you can use a preconfigured Docker container with Python and all the necessary libraries already installed.

1. **Make sure that Docker and Docker Compose** are installed on your system. You can check the installation by running:

```
docker --version
docker-compose --version
```

2. **Navigate to the directory** where the Dockerfile and docker-compose.yml are located. Create the Docker container and execute it:

```
docker-compose up --build
```

This command will create the container (if it has not yet been created) and start the container with Jupyter Lab in it.

3. **Call Jupyter Lab in your browser**:

```
http://localhost:8888/?token=ida2025
```

4. **Stop the container when you finished your work**:

```
docker-compose down
```

This approach with Docker ensures a consistent and portable environment for running Jupyter notebooks and avoids dependency conflicts and installation issues.

4.2.3 LBD Jupyter Python notebook implementations

To ensure the reproducibility of some of the methods and results presented in this monograph, we have implemented several text processing components and LBD pipelines in the form of Jupyter Python notebooks. The tutorials are organized as a series of examples that highlight various features of the implemented pipelines. Each tutorial contains a comprehensive description of the functions and steps involved, guiding the user through the entire process and ensuring that they understand each stage and can apply the knowledge effectively in practical scenarios. Below we outline the implemented text processing components, followed by the implemented LBD pipelines, which can be used for experiment repeatability and code reuse.

Data acquisition and text processing components

The following components are available in the following Jupyter Python notebooks:

LBD_01_data_acquisition.ipynb This notebook includes components for acquiring and loading the text data from various sources like text files, CSV files, or APIs. The list of defined functions includes:

```
load_data_from_file
convert_file_to_ascii_encoding
load_data_from_csv
load_data_from_web
load_data_from_api
load_data_from_pubmed
```

LBD_02_data_preprocessing.ipynb This notebook includes functions that handle the preprocessing tasks for text mining like tokenization, stopword removal, stemming, and lemmatization. The list of defined functions includes:

```
do_clean_text
do_remove_stopwords
do_lemmatize
do_stem
do_keep_only_longer_or_equal_length_words
do_remove_non_nouns
do_remove_non_mesh
preprocess
preprocess_docs_dict
```

LBD_03_feature_extraction.ipynb This notebook includes functions that implement various feature extraction techniques for text mining like Bag of Words, TF-IDF, and word embeddings. The list of defined functions includes:

```
create_bag_of_words
create_tfidf
create_word_embeddings
```

LBD_04_text_mining.ipynb This notebook includes functions that implement various text mining algorithms like topic modeling, sentiment analysis, and clustering. The list of defined functions includes:

```
perform_topic_modeling
perform_sentiment_analysis
perform_clustering
```

LBD_05_results_analysis.ipynb This notebook includes functions that help analyze the results from the text mining algorithms and provide insights into the data. The list of defined functions includes:

```
analyze_topics
analyze_sentiment
analyze_clusters
compute_similarity
find_most_similar_documents
```

LBD_06_visualisation.ipynb This notebook provides visualization functions to help users better understand the results. The list of defined functions includes:

```
plot_wordcloud
plot_sentiment_histogram
plot_clusters_2d
plot_bow_tfidf_matrix
visualize_tfidf_pca_interactive
```

Selected LBD tutorials in Python

Selected LBD pipelines that illustrate how individual LBD methods can be used in practice are implemented in the following Jupyter Python notebooks, summarized in Table 4.3.

Table 4.3 Summary of reproduced LBD techniques.

Technique	Section	Type of Discovery	Dataset	Document Representation
Swanson's ABC (Swanson, 1986)	5.1	Closed	DFO-RS	TF-IDF BoW
Concept-based (Weeber et al., 2001)	7.2	Open	Mig-Mg	BoW
CrossBee (Juršič et al., 2012)	5.3	Closed	AUT-CaN	TF-IDF BoW
OntoGen-based (Cestnik et al., 2016)	6.8	Closed	AUT-CaN	TF-IDF BoW
RaJoLink (Petrič et al., 2012)	7.3	Open	AUT-CaN	TF-IDF BoW
Network-based (Kastrin et al., 2014)	8.2	Closed	DFO-RS	Network

01_closed_discovery.ipynb The workflows that demonstrate a Swanson's ABC approach to LBD, presented in Section 5.1 of Chapter 5, are described in Section 5.3.1.

02_open_discovery.ipynb The workflows that demonstrate a concept mapping approach to semantic concept discovery, presented in Section 7.2 of Chapter 7, are described in Section 7.4.1.

03_mini_crossbee.ipynb The workflows that demonstrate selected functionalities of the CrossBee text mining system for b-term discovery, presented in Chapter 5, are described in Section 5.3.

04_mini_ontogen.ipynb The workflows that demonstrate selected functionalities of the OntoGen text clustering system used for outlier document detection in a closed LBD setting, presented in Chapter 6, are described in Section 6.8.

05_mini_rajolink.ipynb The workflows that demonstrate selected functionalities of the RaJoLink open discovery system, presented in Section 7.3 of Chapter 7, are described in Section 7.4.2.

06_link_pred.ipynb The workflows that demonstrate link prediction functionalities for network-based closed discovery, presented in Section 8.2 of Chapter 8, are described in Section 8.4.

07_find_genes.ipynb The workflows present the embeddings-based relational bisociation discovery approach, presented in Section 9.2 of Chapter 9, are described in Section 9.4.

References

Cestnik, B., Fabbretti, E., Gubiani, D., Lavrač, N., & Urbančič, T. (2016). Exploring outlier documents to investigate potential links between gut microbiota and Alzheimer's disease. *International Work-Conference on Bioinformatics and Biomedical Engineering, IWBBIO 2016*, 475–486.

Juršič, M., Cestnik, B., Urbančič, T., & Lavrač, N. (2012). Cross-domain literature mining: Finding bridging concepts with CrossBee. *Proceedings of the 3rd International Conference on Computational Creativity*, 33–40.

Kastrin, A., Rindflesch, T. C., & Hristovski, D. (2014). Link prediction on the semantic MEDLINE network. *Proceedings of the 17th International Conference on Discovery Science*, 135–143.

Moncada, S. (1983). Biology and therapeutic potential of prostacyclin. *Stroke*, *14*(2), 157–168.

Moncada, S., & Vane, J. (1984). Prostacyclin and its clinical applications. *Annals of Clinical Research*, *16*(5-6), 241–252.

Pedregosa, F., Varoquaux, G., Gramfort, A., Michel, V., Thirion, B., Grisel, O., Blondel, M., Prettenhofer, P., Weiss, R., Dubourg, V., Vanderplas, J., Passos, A., Cournapeau, D., Brucher, M., Perrot, M., Duchesnay, E., & Louppe, G. (2011). Scikit-learn: Machine learning in Python. *Journal of Machine Learning Research*, *12*, 2825–2830.

Peng, R. D. (2011). Reproducible research in computational science. *Science*, *334*(6060), 1226–1227.

Petrič, I., Cestnik, B., Lavrač, N., & Urbančič, T. (2012). Outlier detection in cross-context link discovery for creative literature mining. *The Computer Journal*, *55*(1), 47–61.

Petrič, I., Urbančič, T., Cestnik, B., & Macedoni-Lukšič, M. (2009). Literature mining method RaJoLink for uncovering relations between biomedical concepts. *Journal of Biomedical Informatics*, *42*(2), 219–227.

Re:Science. (2025). Re:Science [Accessed on February 07, 2025]. https://rescience.github.io/

Swanson, D. R. (1986). Fish oil, Raynaud's syndrome, and undiscovered public knowledge. *Perspectives in Biology and Medicine*, *30*(1), 7–18.

Swanson, D. R. (1988). Migraine and magnesium: Eleven neglected connections. *Perspectives in Biology and Medicine*, *78*(1), 526–557.

Toomey, D. (2017). *Jupyter for data science: Exploratory analysis, statistical modeling, machine learning, and data visualization with Jupyter*. Packt Publishing.

Vosgerau, H. (1973). Migraine therapy with magnesium glutamate. *Therapie der Gegenwart*, *112*(4), 640.

Weeber, M., Klein, H., de Jong-van den Berg, L., Vos, R., et al. (2001). Using concepts in literature-based discovery: Simulating Swanson's Raynaud–fish oil and migraine–magnesium discoveries. *Journal of the American Society for Information Science and Technology*, *52*(7), 548–557.

Wolf, T., & et al. (2020, October). Transformers: State-of-the-art natural language processing. In Q. Liu & D. Schlangen (Eds.), *Proceedings of the 2020 Conference on Empirical Methods in Natural Language Processing: System Demonstrations* (pp. 38–45). Association for Computational Linguistics.

Chapter 5
Text Mining for Closed Discovery

This chapter explores literature-based discovery using text mining. Section 5.1 outlines the early Swanson's approach to literature-based discovery, Section 5.2 presents the CrossBee literature-based discovery methodology, by illustrating its ensemble-based bridging term (b-term) ranking and exploration approach. The chapter concludes with Section 5.3, which includes a Python tutorial allowing for simplified LBD experiment replicability and code reuse.

5.1 Undiscovered public knowledge: A tribute to Don R. Swanson

Don R. Swanson was an influential information scientist who believed that scientific discoveries could come not only from laboratory experiments but also from the exploration of previously published scientific literature. He coined the term "undiscovered public knowledge" to describe hidden connections and information in scientific articles that had not yet been identified or exploited. This section presents Swanson's early work on literature-based discovery in Section 5.1.1, outlines the Arrowsmith LBD system in Section 5.1.2, followed by Section 5.1.3 focusing on replicating Swanson's discovery of the link between Raynaud's disease and fish oil.

5.1.1 LBD through serendipity

When examining more than 4,000 Medline article titles, Swanson (1986a) discovered that dietary fish oil (DFO) can reduce blood viscosity, decrease platelet aggregation, and inhibit vascular reactivity, specifically vascular constriction. He observed that reducing both blood viscosity and platelet aggregation, as well as suppressing vasoconstriction, appeared to prevent Raynaud's disease. Raynaud's disease, also known as Raynaud's syndrome or phenomenon in the older literature, is a clinical condition in which the spasm of small arteries causes episodes of reduced blood flow to the end arterioles (usually to fingers and toes, but less commonly to the nose, ears, lips, or nipples).

N. Lavrač et al., *Bisociative Literature-Based Discovery*,
https://doi.org/10.1007/978-3-031-96863-1_5

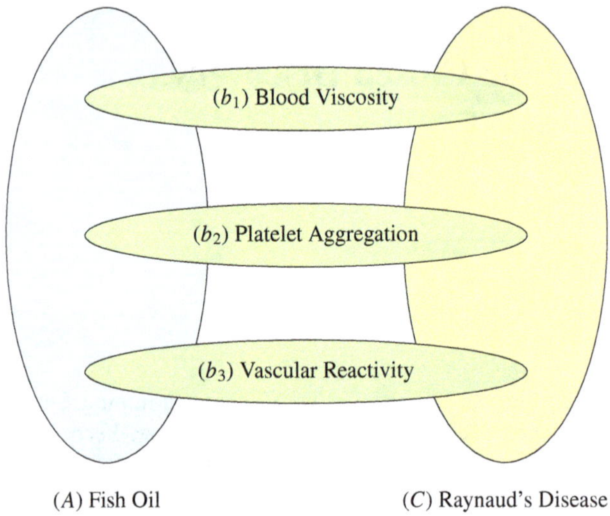

Fig. 5.1 Venn diagram demonstrating Swanson's dietary fish oil–Raynaud's syndrome hypothesis. Fish oil is beneficial for patients with Raynaud's disease through three distinct mechanisms.

Swanson postulated that the consumption of fish oil in one's diet could alleviate or prevent Raynaud's disease. This hypothesis, illustrated in Figure 5.1, is known as the dietary fish oil–Raynaud's syndrome (DFORS) hypothesis.

The guiding idea behind Swanson's work is that the two originally unconnected parts of the literature (C and A) can establish latent connections (through intermediate terms in B). These two pieces of literature have one or more b-terms in common, creating an indirect connection. Once these bridges are identified and validated, new knowledge can be generated.

The *serendipity* of Swanson's discovery refers to the fact that there was no previously published evidence that combined the ideas of "dietary fish oil" and "Raynaud's syndrome" in a single document. The novelty of Swanson's finding can be attributed to the observation that the existing literature had already established connections between "dietary fish oil" and intermediate concepts such as blood viscosity, platelet aggregation, and vasoconstriction (e.g., Bang et al., 1971). On the other hand, explicit links between these intermediates and Raynaud's disease were also well documented (e.g., Pringle et al., 1965).

Swanson founded his discovery on the so-called ABC model, which is now widely regarded as a typical paradigm for LBD. Recall from Chapter 1 that label C denotes the source concept, label A stands for the target concept, and label B denotes intermediate concepts that are yet to be discovered. In Swanson's scenario, label C refers to Raynaud's disease concept (set of papers mentioning Raynaud's disease), whereas label A refers to the fish oil concept (set of papers mentioning fish oil). At the end of November 1985 when Swanson executed his experiments, there were approx. 1,000 papers in the Raynaud's disease set (C), and 3,000 papers in the fish oil set (A).

Let us examine Swanson's results more closely. When inspecting the articles on Raynaud's disease (*C*), Swanson noted that patients suffering from Raynaud's disease exhibited high values of some hematological parameters, including blood viscosity and platelet aggregation. In response to low temperatures, these patients are also characterized by vascular reactions such as increased vasoconstriction. These features are the so-called *b*-terms (see terms b_1, b_2 and b_3 in Figure 5.1), which form explicit *BC* knowledge in the context of Raynaud's disease (*BC* refers to concepts/terms in the intersection of literatures $B \cap C$). On the other hand, Swanson recognized that the same *b*-terms were also reported in papers related to fish oil (*A*). Specifically, in reviewing the *AB* literature for concepts/terms in the intersection of literatures $A \cap B$, he highlighted that eicosapentaenoic acid, an omega-3 polyunsaturated fatty acid present in cold-water fish, lowers blood viscosity and platelet aggregation. Additionally, fish oil promotes vasodilation (i.e. it inhibits vasoconstriction). Although the relations between concepts *A* and *B* and, likewise, between *B* and *C* were described in the literature much before Swanson's experiment in 1985, Swanson was the first to establish a hypothetical link between concepts *A* and *C*. Three years after Swanson's initial discovery, a clinical trial confirmed the use of fish oil in the treatment of Raynaud's disease, providing evidence in support of his DFORS hypothesis (DiGiacomo et al., 1989).

Empirical findings on the DFORS hypothesis presented by Swanson (1986a) were complemented by a theoretical discussion in the article "Undiscovered Public Knowledge" in which Swanson (1986b) discussed connecting unrelated research to uncover new information. In his own words:

> Imagine that the pieces of a puzzle are independently designed and created, and that, when retrieved and assembled, they then reveal a pattern – undesigned, unintended, and never before seen, yet a pattern that commands interest and invites interpretation. So it is, I claim, that independently created pieces of knowledge can harbor an unseen, unknown, and unintended pattern. And so it is that the world of recorded knowledge can yield genuinely new discoveries (Swanson, 1986b).

Swanson later conducted several other case studies that illustrated various disease treatment combinations in which two separate bodies of literature contained complementary knowledge:

- migraine disorders and magnesium (Swanson, 1988)
- somatomedin C and arginine (Swanson, 1990b),
- magnesium deficiency and neurologic disease (Smalheiser & Swanson, 1994),
- Alzheimer's disease and indomethacin (Smalheiser & Swanson, 1996a),
- Alzheimer's disease and estrogen (Smalheiser & Swanson, 1996b), and
- schizophrenia and calcium-independent phospholipase A2 (Smalheiser & Swanson, 1998).

The impact of Swanson's work extended beyond the specific discovery of fish oil for Raynaud's disease. Swanson's work on LBD and his concept of undiscovered public knowledge have had a significant impact on the field of information science. In 2000, he received the ASIS&T Award of Merit, the highest honor from the American Society for Information Science & Technology.[1]

[1] The Association for Information Science and Technology (ASIS&T) is the leading professional organization bridging the gap between the practical application and research of information science.

5.1.2 Arrowsmith LBD tool

A significant disadvantage of the manual LBD approach is that it requires a large number of manual interventions from the user. Generating a list of interpretable b-terms on a trial-and-error basis is a tedious and time-consuming process. In the mid-1990s, Swanson and Smalheiser initiated a project to reduce the amount of human intervention in the discovery process as much as possible. A decade after the manual discovery of the DFORS hypothesis, they presented the first LBD tool, called Arrowsmith (Swanson & Smalheiser, 1997). At the time of this writing, the scientific community recognizes the Arrowsmith tool as the most popular LBD system with a well-maintained Web application that is easy to use.[2] A screenshot showing the use of the Arrowsmith LBD tool is shown in Figure 5.2.

Arrowsmith was specifically developed to aid biomedical researchers in the process of building and validating scientific hypotheses. Its broader functionality enables users to uncover biologically meaningful connections between two disjoint sets of papers, previously denoted as A and C, within PubMed. This capability extends even to scenarios where there are no overlapping articles or authors between sets A and C, representing divergent topics or disciplines.

5.1.3 Replicating the early discovery by Swanson

Swanson first experimented with a one-node search, in which he started with a single body of literature and searched for a second body of complementary knowledge. However, his main methodological framework is a two-node search approach that is more generally applicable: the user specifies two disjoint sets of literatures (A and C) that are assumed to share some relevant latent links.

We illustrate the general principle of utilizing Arrowsmith to produce a list of *b*-terms, with the goal of replicating Swanson's (1986a) findings related to the DFORS hypothesis. In this experiment, only titles of papers are used. We have organized this discovery process into six sequential steps:

1. Identify the concepts of interest, i.e. the source concept c and the target concept a. In the case of the DFORS hypothesis, the phrase "raynaud" denotes the concept c, and the phrase "fish oil" represents the concept a.
2. Retrieve all PubMed papers that contain the concept a. To simulate Swanson's settings, we limit the results to November 30, 1985, using the PubMed query `raynaud[TI] AND 1960/01/01:1985/11/30[PDAT]`.

ASIS&T connects researchers from diverse fields, including information science, computer science, linguistics, librarianship, engineering, data science, and related technologies. Established in 1964, the ASIS&T Award of Merit is a prestigious honor presented annually to an individual who has made a notable contribution to the field of information science. Such contributions may include the introduction of new ideas, the creation of innovative devices, the development of improved techniques, or outstanding service to the information science profession.

[2] https://arrowsmith.psych.uic.edu

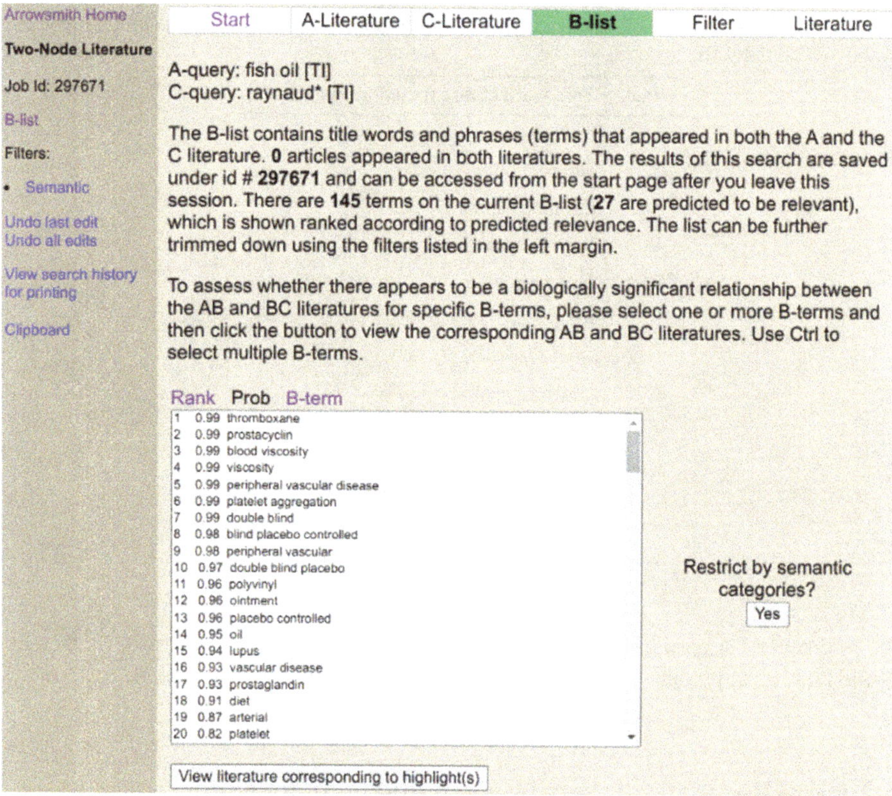

Fig. 5.2 Screenshot of the B-list page from the Arrowsmith LBD tool, where the b-terms are listed in a text box and ordered by their predicted relevance.

3. Similarly, we collect all the papers with the concept c: `fish oil[TI] AND 1960/01/01:1985/11/30[PDAT]`.
4. Remove papers present in both sets A and C to consider only indirect connections between the sets A and C. Arrowsmith executes this step automatically, without the user's intervention, during the preprocessing stage.
5. Find the b-terms that are in set A and in set C. The resulting list of bridging terms is generated from terms (i.e., individual words, bi-grams or three-grams) extracted from the titles of papers in both sets A and C.
6. Generate AB and BC lists. For each b-term b_i, Arrowsmith generates an AB list of titles containing both "fish oil" and b_i, and a BC list of titles containing both b_i and "raynaud"; these titles are juxtaposed to facilitate the user judging whether there is likely to be a biologically relevant relation among "fish oil", set of terms B, and "Raynaud's diseas".

The resulting list of b-terms is shown in Table 5.1. These b-terms are concepts that appear as words or phrases in the titles of papers in A and C literatures. The candidate b-terms are ranked based on their b-term potential, using an ensemble heuristic, combining

Table 5.1 The list of title words and phrases that form b-terms across literatures A and C, suggested by the Arrowsmith LBD system.

Rank	Title word/phrase (potential b-terms)
1	prostacyclin
2	deformability
3	thromboxane
4	lipid lipoprotein
5	*blood viscosity*
6	peripheral vascular disease
7	*vascular reactivity*
8	platelet function
9	peripheral vascular
10	*platelet aggregation*
11	viscosity
12	lipid lipoprotein patient

the estimates of several individual heuristics using a linear regression approach. Details of the ranking approach can be checked in the Arrowsmith documentation.

5.2 CrossBee methodology for bridging term detection

In cross-domain knowledge discovery, estimating which terms have a high potential for interesting discoveries is a challenging research question. It is especially important for cross-context scientific discovery such as understanding complex medical phenomena or finding new drugs for yet poorly understood diseases. This section focuses on the CrossBee methodology for bridging term detection.[3]

5.2.1 Heuristics for bridging term discovery

Different elementary and ensemble heuristics used for b-term ranking are available in CrossBee. Heuristics are defined as functions that numerically evaluate the term's quality by assigning a *bisociation score* to a term (measuring the potential that a term is actually a b-term). For the definition of an appropriate set of heuristics, we define a set of special (mainly statistical) properties of terms, which aim at distinguishing b-terms from regular terms; thus, these heuristics can also be viewed as advanced term statistics.

Formally, a heuristic is a function with two inputs, i.e. a set of domain-labeled documents D and a term t appearing in these documents, and one output, i.e. a score that

[3] This section is largely based on the papers by Juršič et al. (2012a) and Juršič et al. (2012b).

represents the term's bisociation potential. All heuristics operate on the data retrieved from the documents in text preprocessing. Although term ranking using the scores calculated by an ideal heuristic should result in ranking all b-terms at the top of the ranked list, this ideal scenario is not realistic; nevertheless, ranking by heuristic scores should at least increase the proportion of b-terms at the top of the ranked term list.

We use the following notation: to state that the term's bisociation score b is equal to the result of a heuristic named $heurX$, we can denote it as $b = heurX(D, t)$. However, since the set of input documents is static when dealing with a concrete dataset, we can—for the sake of simplicity—omit the set of input documents from a heuristic notation and use only $b = heurX(t)$. Whenever we need to explicitly specify the set of documents to which the function is applied (never needed for a heuristic, but sometimes needed for auxiliary functions used in the formula for the heuristic), we write it as $heurX_D(t)$. For specifying the function's input document set, we have two options: either use D_u that stands for the (union) set of all the documents from all the domains, or use $D_n : n \in \{1..N\}$, which stands for the set of documents from the given domain n. In general, the following statement holds: $D_u = \cup_{n=1}^{N} D_n$ where N is the number of domains. In the most common scenario, when there are exactly two distinct domains, we also use the notation D_A for D_1 and D_C for D_2, similar to Swanson's notation of symbols A and C as representatives of the target and the initial domains in the closed discovery setting, introduced in Section **??**.

Elementary heuristics. Four sets of elementary heuristics are implemented in CrossBee: six frequency-based, four TF-IDF weights-based, three similarity-based, and eight outlier-based heuristics. The first two groups are motivated by the idea that frequent terms in the intersections of documents of distinct domains A and C may be useful for cross-domain discovery, as terms b frequently mentioned in each of the two domains may represent some phenomena related to individual concepts a and c, until now being separately explored in research literatures A and C. The third group is motivated by trying to explore contextual similarities: similarity of contexts b mentioned in literature A with contexts in which b is mentioned in literature C. The fourth group is motivated by exploring potential b-terms in outlier documents.

- The heuristics in the first three groups work fundamentally in a similar way: they only manipulate the data present in the Bag-of-Words (BoW) document vector format to derive the term bisociation potential quality measure, named the *bisociation score*. Among them, the TF-IDF heuristic (introduced in Equation 3.1) is the most frequently used heuristic in the BoW vector construction, where TF-IDF denotes the product of Term Frequency and Inverse Document Frequency weights, frequently used in document vector representations used in text mining (Salton & Buckley, 1988).
- The exceptions are outlier-based heuristics, forming the fourth group of heuristics, which first detect outlier documents and then use the information from BoW vectors. The rationale for considering outlier-based heuristics is explained in Chapter 6, devoted to outlier-based b-term detection.

The main novelty of the fourth group of heuristics is the consideration of b-term frequencies in sets of outlier documents, motivated by the observation that documents in the outliers set frequently embody new information that is often hard to explain in the context of existing knowledge. In line with the research outlined in Chapter 6, we concentrated on a specific type of document outliers, i.e. domain outliers, which refer to documents that tend to be more similar to the documents of the other domain A than to those of their own domain C, and are therefore misclassified into a different class than their original class.

The argument why one can focus on searching for b-terms in outlier documents identified by classification algorithms is explained in Section 6.2. The procedure we use to detect outlier documents first builds a classification model for each domain and then classifies all documents using the trained classifier. The documents that are misclassified are declared as outliers since according to the classification model they do not belong to their initial domain. The other two outlier-based heuristics—relative frequency in the outlier set detected by the Random Forest (RF) classifier *outFreqRelRF* and the relative frequency in the outlier set detected by the Support Vector Machine (SVM) classifier *outFreqRelSVM*—are defined in the same way as the *outFreqRelCS* heuristic.

For illustrative purposes, we specify a set of selected elementary heuristics whose performance was tested for their bisociative potential in the CrossBee system. The selected heuristics are outlined as follows.

outFreqRelRF $= \frac{countTerm_{D_{RF}}(t)}{countTerm_{D_u}(t)}$: This heuristic computes the relative frequency of term t in the outlier document set detected by the Random Forest classifier,

outFreqRelSVM $= \frac{countTerm_{D_{SVM}}(t)}{countTerm_{D_u}(t)}$: This heuristic computes the relative frequency of term t in outlier document set detected by the Support Vector Machine classifier,

outFreqRelCS $= \frac{countTerm_{D_{CS}}(t)}{countTerm_{D_u}(t)}$: This heuristic computes the relative frequency of term t in the outlier document set detected by the Centroid Similarity classifier,

outFreqSum $= countTerm_{D_{CS}}(t) + countTerm_{D_{RF}}(t) + countTerm_{D_{SVM}}(t)$: This heuristic computes the sum of frequencies of term t in all three outlier document sets,

tfidfDomnSum $= tfidf_{D_1}(t) + tfidf_{D_2}(t)$: This heuristic computes the sum of term TF-IDF weights of term t in the two domains,

freqRatio $= \frac{countTerm_{D_u}(t)}{countDoc_{D_u}(t))}$: This heuristic computes the term to document frequency ratio.

We have defined also a supplementary baseline heuristic: *random(t) = randNum()* that served as a baseline for the others, as it returns a random number in the interval $(0, 1)$ regardless of the term under investigation.

Using these base heuristics, we developed the ensemble heuristic described in Section 5.2.2 below. A detailed justification for the choice of this particular combination of heuristics is presented in Juršič (2015).

Evaluation of elementary heuristics. We used ROC analysis to test the b-term discovery potential of various CrossBee heuristics. We compared the above six heuristics by depicting their ROC curves and comparing their AUC (Area Under ROC) values. Unlike in standard ROC analysis (see Section 3.7.2 for details), the idea underlying ROC curve construction in this study was to see how well a particular heuristic ranks the 43 ground-truth b-terms, known from the original Swanson's migraine-magnesium study.

Among the terms extracted from the titles of a subset of PubMed titles of articles published before 1988 (i.e. before Swanson's literature-based discovery of the migraine-magnesium relation, and were retrieved with the PubMed search engine, consisting of 2,425 migraine and 5,633 magnesium PubMed articles' titles), there were also all 43 terms that Swanson marked as b-terms and which we hoped to propagate to the top of the ranked list using the designed heuristics. The b-terms identified by Swanson, verified by the expert to provide new discoveries in the field, were used as a gold standard in the evaluation of this work. These 43 terms appear on the y-axis of the plot shown in Figure 5.3, while the x-axix corresponds to all extracted terms.

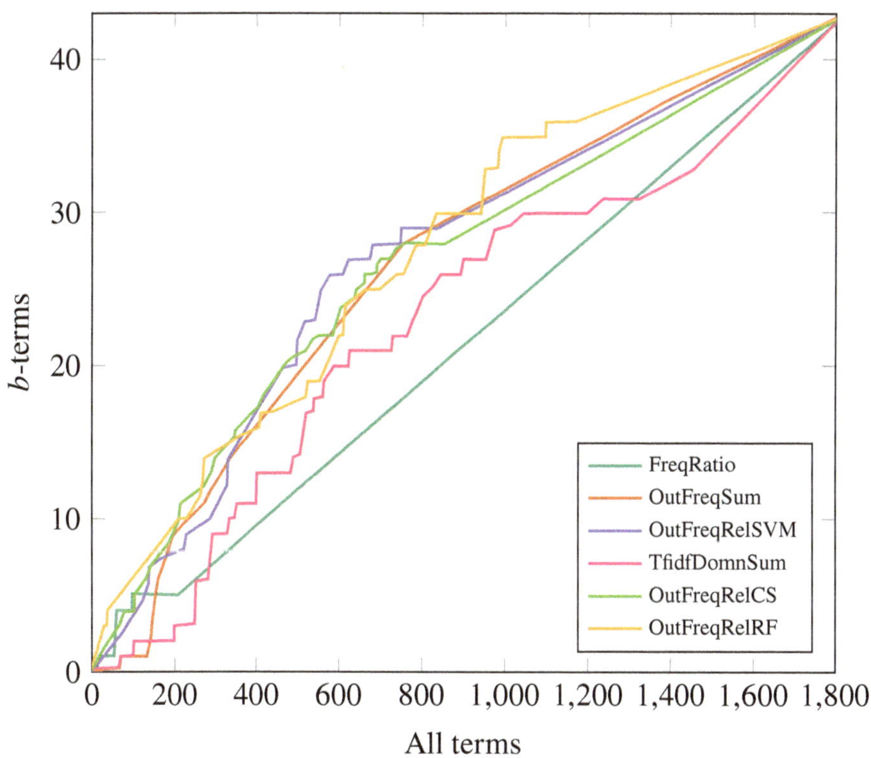

Fig. 5.3 ROC curves representing the performance of six heuristics on the migraine-magnesium training dataset, where the y-axis refers to the 43 terms that Swanson marked as b-terms in his original study in 1988.

For this assessment, ROC curve construction was performed as follows: starting from the beginning of a ranked list, every time a ground-truth b-term is encountered in the list of potential b-terms suggested by a given heuristic, draw a line up on the ROC canvas,

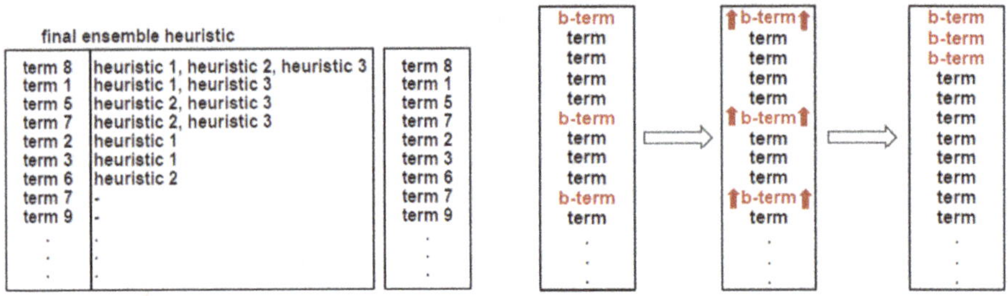

Fig. 5.4 Term ranking approach: first, ensemble heuristics vote for terms, next, terms are sorted according to their potential b-term (as shown on left). Consequently, bridging terms with the highest bridging term potential should receive the highest scores (as shown on the right side).

otherwise draw a line right. Figure 5.3 illustrates the use of ROC analysis assessment of the six selected heuristics in the migraine-magnesium gold standard dataset. For comparison, the ideal curve (if all b-terms were listed at the very beginning of a ranked list proposed by a given heuristic) would go straight up to the top-left angle followed by a straight right line towards the top-right angle of the ROC space; in this case, the area under the ideal ROC curve would be equal to 1 if both scales were normalized.

We observe that some heuristics are very well designed to support b-term discovery. The heuristics that show good performance at the start of the ranked list are especially useful, e.g., heuristic *outFreqRelRF* that places four b-terms already among the first 50 terms in its ranked list, while the random approach finds less than one b-term among its first 200 terms. On the other hand, some heuristics do not perform well at the start of the list, e.g., *outFreqSum* and *tfidfDomnSum*, which prove not being useful for b-term discovery. Nevertheless, we included them in the set of six heuristics forming the ensemble based on their complementarity, so that they fit together well when used in the CrossBee's ensemble heuristic, providing not only better performance but also greater robustness of the ensemble. For details, see Section 5.2.2 below.

5.2.2 Ensemble heuristics

Instead of choosing an individual heuristic for bridging term detection, Juršič et al. (2012a) proposed using an ensemble heuristic. The calculation of the ensemble score has been well justified by an extensive experimental evaluation (Juršič et al., 2012a; Juršič et al., 2012b) on the migraine-magnesium dataset (Swanson, 1988). Based on the experimental results, the final set of elementary heuristics included in the ensemble consisted of the heuristics outlined in Section 5.2.1 above. The ensemble-based ranking methodology for bridging term (b-term) detection is illustrated in Figure 5.4 showing term ranking using an ensemble heuristic.

The ensemble heuristic for bridging term discovery, which was constructed based on the experiments, is computed as a sum of two parts: $s_t = s_t^{vote} + s_t^{pos}$, i.e. the ensemble voting score s_t^{vote} and the ensemble position score s_t^{pos}, which are summed together to give the final ensemble score for every term in the corpus vocabulary. Each term score represents the term's potential for linking the two disjointed domains.

The ensemble voting score (s_t^{vote}) of a given term t is an integer, which denotes how many base heuristics voted for the term: each term can get a score $s_{t_j}^{vote} \in \{0, 1, 2, ..., k\}$, where k is the number of base heuristics used in the ensemble. The ensemble voting score of term t_j that is at position p_j in the ranked list of n terms is computed as a sum of individual heuristics' voting scores:

$$s_{t_j}^{vote} = \sum_{i=1}^{k} s_{t_j,h_i}^{vote} = \sum_{i=1}^{k} \begin{cases} 1, & p_j < n/3 \\ 0, & \text{otherwise.} \end{cases} \tag{5.1}$$

The ensemble position score (s_t^{pos}) is calculated as an average of position scores of individual base heuristics. For each heuristic h_i, the term's position score s_{t_j,h_i}^{pos} is calculated as $\frac{n-p_j}{n}$, which results in position scores being in the interval $[0, 1)$. For an ensemble of k heuristics, the ensemble position score is computed as an average of individual heuristics' position scores:

$$s_{t_j}^{pos} = \frac{1}{k} \sum_{i=1}^{k} s_{t_j,h_i}^{pos} = \frac{1}{k} \sum_{i=1}^{k} \frac{n - p_j}{n}. \tag{5.2}$$

Altogether, the CrossBee's application potential was tested on three LBD problems. To explore the utility of the system, CrossBee was first applied to the migraine–magnesium dataset (Swanson, 1988, 1990a). The second dataset, which served as a test case to demonstrate the domain independence of the methodology, was the autism–calcineurin domain pair introduced and examined by Urbančič et al. (2007) and Petrič et al. (2009). In addition, the third area for searching for bisociative links between literature was the gut microbiome–Alzheimer's disease dataset, described in more detail in Section 6.6.

Like in the experiments in which we evaluated the quality of individual heuristics (see Section 5.2.1), we used the standard migraine-magnesium dataset to evaluate the ensemble heuristic. To mimic Swanson's original study of the migraine literature (Swanson, 1988), we performed the experiments on a subset of PubMed titles of articles that were published before 1988 when retrieved with the PubMed search engine. As a result, we obtained 2,425 migraine and 5,633 magnesium PubMed articles titles. These article titles were preprocessed with standard text mining techniques, resulting in over 13,000 distinct terms, which were analyzed and scored with the CrossBee ensemble heuristic. After each heuristic assigned a score to every term from the list, we sorted all 7 lists (6 elementary heuristics and the baseline heuristic) and in this way created 7 ranked lists of terms. Among these potential b-terms, there were also all 43 terms that Swanson marked as b-terms and which we hoped to propagate to the top of the ranked list using the designed heuristics methodology.

Figure 5.5 shows the list of the top b-terms ranked by CrossBee's ensemble voting heuristics, where the presented ranked list is the actual output produced by the CrossBee bridging term exploration system using the migraine-magnesium gold standard dataset (Swanson, 1988).

B-Term Identify (Analysis)

List start position: 0 Search terms(?): GO

There are 8058 documents in the database with 13445 terms (the termwhitelist contained 0 terms).
You have provided 43 bterms. Out of them 43 are found in the documents.

Pos.	Term	Votes	Inner Class Score	Documents	
				MIG	MAG
1	clinical		0,9935	115	88
2	therapy		0,9928	105	108
3	treatment		0,9923	362	156
4	trial		0,9915	73	16
5	case		0,9910	79	55
6	patient		0,9902	180	288
7	test		0,9902	17	37
8	syndrome		0,9901	45	44
9	magnesium		0,9899	1	5628
10	cerebral		0,9898	78	25
11	control		0,9898	68	70
12	drug		0,9896	75	32
13	pain		0,9894	33	5
14	study		0,9892	187	375
15	serotonin [1]		0,9892	63	8
16	artery		0,9891	41	24
17	prevention		0,9886	49	26
18	disease		0,9885	23	174
19	blood		0,9884	71	235
20	acid		0,9884	45	201

Fig. 5.5 The ensemble heuristic based ranking page showing the top 20 ranked terms from the migraine-magnesium domain.

The reader can find Python tutorials on the use of the explained CrossBee ensemble-based LBD approach, applied in two experimental scenarios (migraine-magnesium and autism-calcineurin), in Section 5.3.

5.2.3 CrossBee's HCI interface for bridging term discovery

The user-friendly CrossBee Web interface was designed to efficiently investigate cross-domain links ranked by the ensemble-based ranking methodology, and to inspect the actual appearances of the selected b-term in documents of both domains. Although at the time of writing the monograph the Web interface is no longer functional, we include the screenshots for documentation and to illustrate good HCI design principles used in the system's Web interface conception.

The ranked list of b-term candidates, shown in Figure 5.5, provides the user with additional information including the term domain occurrence statistics in both domains.

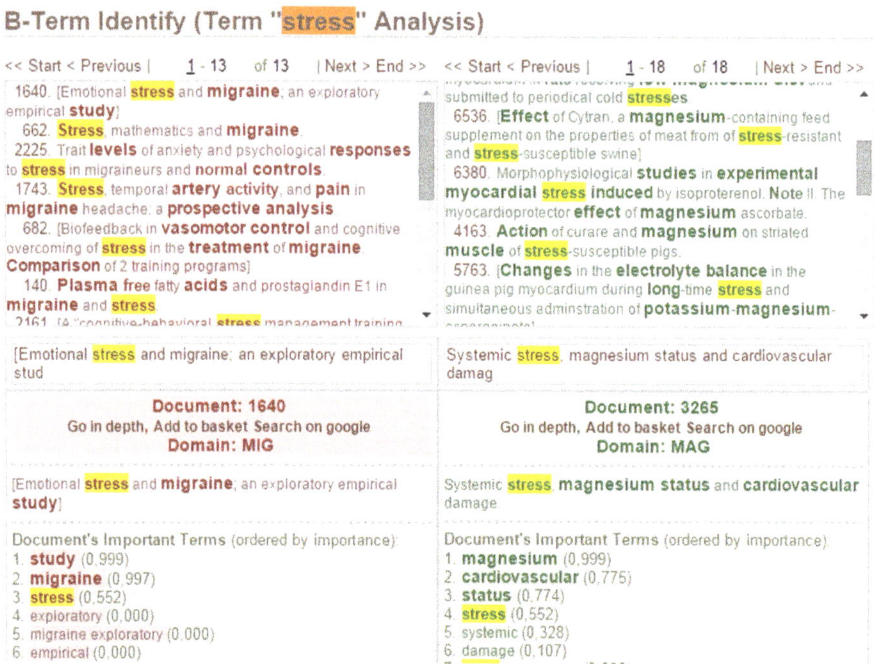

Fig. 5.6 One of the useful features of the CrossBee interface was the side-by-side view of documents from the two domains under investigation. The analysis of the "stress" term from the migraine–magnesium domain is shown. The presented view enables an efficient comparison of two documents, the left one from the migraine domain and the right one from the magnesium domain.

Using the HCI interface, the user can also inspect the actual appearance of the selected term in documents of both domains, using the side-by-side document inspection as shown in Figure 5.6. In this way, the user can verify whether the rationale behind selecting a given term as a bridging term can be justified based on the contents of the inspected documents.

The CrossBee's document-focused exploration interface, illustrated in Figure 5.6, empowers the user when filtering and ordering the documents by various criteria, including a detailed document view that provides a more detailed presentation of a single document including various term statistics. Methodology performance analysis supports the evaluation of the methodology by providing various data that can be used to measure the quality of the results, e.g., data for plotting the ROC curves. High-ranked terms emphasis is used to mark the terms according to their bisociation score calculated by the ensemble heuristic. When using this feature, all the high-ranked terms are emphasized throughout the entire application, thus making them easier to spot (see different font sizes in Figure 5.6). b-term emphasis marks the terms defined as b-terms by the user (yellow terms in Figure 5.6). Domain separation is a simple but effective option that colors all documents from the same domain with the same color, making an obvious distinction between the documents from the two domains (different colors in Figure 5.6). User interface customization enables the user to decrease or increase the intensity of the following features: high-ranked term emphasis, b-term emphasis, and domain separation.

In summary, the CrossBee system was designed to suggest bridging terms between two user-defined domains (separate domain literatures) using an ensemble-based term ranking method. The tool also helped experts explore the hidden links that connect two seemingly unrelated domains. In addition to this core functionality, other features and content representations have been added to make the CrossBee Web application support the user in exploring possible cross-context links. This allowed the user not only to identify discovered cross-domain links, but also to explore them efficiently and effectively.

5.3 Tutorials in Python

The content presented in this chapter is supplemented by a practical example whose source code is hosted on GitHub. We recommend that readers access the tutorials **01_closed_discovery** and **03_mini_crossbee** in the book's GitHub repository to further explore the concepts discussed. Before running the Jupyter notebooks, the readers are advised to read Chapter 4 and follow the instructions to install Python and the required packages from the **requirements.txt** file in the repository.

5.3.1 01_closed_discovery.ipynb

This notebook illustrates the general principle of LBD based on Swanson's (1986a) the ABC closed discovery model, where A and C refer to the two domains in the A–C domain pair, and B refers to a set of potential b-terms. The goal is to replicate the link between Raynaud's syndrome and dietary fish oil (i.e. the RS-DFO hypothesis) employing the Swanson's closed discovery approach.

For the RS-DFO hypothesis, the term "raynaud" represents the concept c, while "fish oil" corresponds to the concept a. To simulate the closed discovery mode, we exclude any records shared between domains C and A, thereby focusing exclusively on indirect connections between the two domains. For document representation, we use article titles, while terms are represented as bigrams. Domain C comprises 4554 unique terms (uni- and bi-grams), while domain A contains 1059 distinct terms. We then identify intersecting terms shared between the two domains. This set consists of 27 common terms, including interesting b-terms such as "blood viscosity", "platelet aggregation", and "vascular reactivity", which were originally identified by Swanson (1986a).

5.3.2 03_mini_crossbee.ipynb

This workflow illustrates the experiments with CrossBee in the autism-calcineurin domain. The Jupyter notebook named `03_mini_crossbee.ipynb` is available in the repository of this monograph: `https://github.com/lbd-book/code/`. Note that by setting the value of the `CONTEXT_SWITCH` variable in the notebook, the reader can also experiment with other preselected domains, like, for example, the migraine–magnesium domain.

The CrossBee method (Juršič et al., 2012a) represents documents using a BoW document representation, where TF-IDF weights are used to model the importance of words and multi-word (n-gram) expressions. The method works on a dataset spanning over two domains, Autism and Calcineurin, with 10,819 documents from domain Autism and 5,320 documents from domain Calcineurin (Table 4.1). For each domain, TF-IDF weighted BoW representations are built. Next, each domain representation is aggregated, allowing for easier analysis of the text data categorized by domain. On top of the calculated statistics, several pre-prepared heuristics are used to estimate the importance of terms as bridging term candidates. These heuristics include elementary heuristics, such as *freqTerm*, *freqDoc*, *freqRatio*, which can be combined in an ensemble heuristic for b-term discovery.

From the 107,261 terms and bigrams of the initial vocabulary, 37,200 terms that occur in both domains are selected as candidates for bridging terms and ranked according to a selected ensemble heuristic. The position of the b-terms from Table 4.2 of Chapter 4 is determined and displayed in the ROC diagram. The area under the curve (AUC) is also calculated for the selected ensemble heuristic.

References

Bang, H., Dyerberg, J., & Nielsen, A. B. (1971). Plasma lipid and lipoprotein pattern in Greenlandic West-coast Eskimos. *The Lancet*, *297*(7710), 1143–1146.

DiGiacomo, R. A., Kremer, J. M., & Shah, D. M. (1989). Fish-oil dietary supplementation in patients with Raynaud's phenomenon: A double-blind, controlled, prospective study. *The American Journal of Medicine*, *86*(2), 158–164.

Juršič, M., Cestnik, B., Urbančič, T., & Lavrač, N. (2012a). Cross-domain literature mining: Finding bridging concepts with CrossBee. *Proceedings of the 3rd International Conference on Computational Creativity*, 33–40.

Juršič, M. (2015). *Text mining for cross-domain knowledge discovery* [Doctoral dissertation, Jožef Stefan International Postgraduate School].

Juršič, M., Cestnik, B., Urbančič, T., & Lavrač, N. (2012b). Bisociative literature mining by ensemble heuristics. In M. R. Berthold (Ed.), *Bisociative knowledge discovery* (pp. 338–358). Springer.

Petrič, I., Urbančič, T., Cestnik, B., & Macedoni-Lukšič, M. (2009). Literature mining method RaJoLink for uncovering relations between biomedical concepts. *Journal of Biomedical Informatics*, *42*(2), 219–227.

Pringle, R., Walder, D., & Weaver, J. (1965). Blood viscosity and Raynaud's disease. *The Lancet*, *285*(7395), 1086–1089.

Salton, G., & Buckley, C. (1988). Term-weighting approaches in automatic text retrieval. *Information Processing and Management*, *24*(5), 513–523.

Smalheiser, N. R., & Swanson, D. R. (1994). Assessing a gap in the biomedical literature: Magnesium deficiency and neurologic disease. *Neuroscience Research Communications*, *15*(1), 1–9.

Smalheiser, N. R., & Swanson, D. R. (1996a). Indomethacin and Alzheimer's disease. *Neurology*, *46*(2), 583–583.

Smalheiser, N. R., & Swanson, D. R. (1996b). Linking estrogen to Alzheimer's disease: An informatics approach. *Neurology*, *47*(3), 809–810.

Smalheiser, N. R., & Swanson, D. R. (1998). Calcium-independent phospholipase a2 and schizophrenia. *Archives of General Psychiatry*, *55*(8), 752–753.

Swanson, D. R. (1986a). Fish oil, Raynaud's syndrome, and undiscovered public knowledge. *Perspectives in Biology and Medicine*, *30*(1), 7–18.

Swanson, D. R. (1986b). Undiscovered public knowledge. *The Library Quarterly*, *56*(2), 103–118.

Swanson, D. R. (1988). Migraine and magnesium: Eleven neglected connections. *Perspectives in Biology and Medicine*, *78*(1), 526–557.

Swanson, D. R. (1990a). Medical literature as a potential source of new knowledge. *Bulletin of the Medical Library Association*, *78*(1), 29–37.

Swanson, D. R. (1990b). Somatomedin C and arginine: Implicit connections between mutually isolated literatures. *Perspectives in Biology and Medicine*, *33*(2), 157–186.

Swanson, D. R., & Smalheiser, N. R. (1997). An interactive system for finding complementary literatures: A stimulus to scientific discovery. *Artificial Intelligence*, *91*(2), 183–203.

Urbančič, T., Petrič, I., Cestnik, B., & Macedoni-Lukšič, M. (2007). Literature mining: Towards better understanding of autism. *Proceedings of the 11th Conference on Artificial Intelligence in Medicine*, 217–226.

Chapter 6
Outlier-based Closed Discovery

This chapter addresses literature-based discovery focused on outlier documents, aiming to improve b-term search efficiency by reducing the search space of potential b-terms to those appearing in outlier documents. Section 6.1 provides a motivation for using outliers in LBD. The next two sections describe two methods for b-term detection in outlier documents. Section 6.2 presents the classification approach for detecting outlier documents, while Section 6.3 outlines a clustering approach to outlier document detection. In Section 6.4, outlier detection is compared to a heuristic approach to b-term detection, as implemented in the CrossBee LBD system. Then, the combined methodology is presented in Section 6.5 as a two-step process combining outlier detection and cross-domain term exploration using the CrossBee tool, illustrated by workflows in the TextFlows Web-based platform. Section 6.6 illustrates the application of the method to the domains of Alzheimer's disease and gut microbiota. Section 6.7 evaluates and compares the obtained results with other research results and provides a summary and directions for further work. The chapter concludes with Section 6.8, which includes a Python tutorial allowing for simplified LBD experiment replicability and code reuse.

6.1 Motivation for using outliers in LBD

In statistics, an outlier is described as an observation that deviates numerically from the rest of the data, or, more formally, an observation that falls out of the overall pattern of a distribution (Moore et al., 2009). While outliers in many data sets may be due to measurement error (such that it would be best to drop them from the data), there are also some examples where outliers have actually led to important discoveries of intriguing information.

Outlier detection is an established area of data mining (Aggarwal, 2013). Conceptually, an outlier is an unexpected event, entity or—in our case—an irregular document. We are especially interested in outlier documents since they frequently embody new information that is often hard to explain in the context of existing domain knowledge. Moreover, in data mining, an outlier is occasionally a primary object of study as it can

N. Lavrač et al., *Bisociative Literature-Based Discovery*,
https://doi.org/10.1007/978-3-031-96863-1_6

potentially lead to the discovery of new knowledge. These assumptions are well aligned with the bisociation potential explored in LBD.

6.2 Document classification for outlier detection and b-term identification

The outlier-based b-term detection methodology is based on research conducted by Petrič et al. (2012a) and Sluban et al. (2012) that has demonstrated that bridging terms are substantially more frequent in documents that are outlier documents of their own domain, compared to their frequency in typical (non-outlier) documents. Similarly to statistics, where an outlier is defined as an observation that falls outside the overall pattern of a distribution (Moore et al., 2009), an outlier document in the field of LBD is a document that lies outside the main group of documents of its own domain and is therefore, in a two-domain *A*-*C* setting, considered more similar to the documents of the other explored domain than to the documents of the domain of its own origin.

The methodology explained in this section exploits the detection of outlier documents to assist the expert in finding implicit links between concepts from two different areas of interest, *A* and *C*. The underlying assumption is that while most articles in the given scientific domain describe facts related to a general understanding of the domain or to more intensively studied domain topics, the exploration of outlier documents can lead to the discovery of scientifically, pharmacologically or clinically relevant bridging concepts between sets of scientific articles from two disjoint domains in a novel, yet unexplored way (Cestnik et al., 2016).

One of the techniques that can be used to detect outliers is the use of classification algorithms (Sluban et al., 2012). The technique works as follows. Having documents from two domains of interest, *A* and *C*, one first trains a classification model that distinguishes between the documents of these domains. The constructed model can then be used to classify all documents (i.e. documents from $A \cup C$). Documents that are misclassified according to their domain of origin (i.e. documents from *A* classified into *C*, and documents from *C* classified as *A*, respectively) are declared outlier documents. These domain outliers are considered borderline documents as they are considered by the model to be more similar to the other domain than to the domain of origin. Sluban et al. (2012) used classification algorithms to find outlier documents, focusing on cross-domain bridging terms between two predefined domains of the researcher's interest, *A* and *C*. The experiments were designed to validate the hypothesis that sets of outlier documents are rich in b-terms and contain significantly more b-terms than sets of arbitrary documents.

When exploring a domain pair dataset, Sluban et al. (2012) searched for a set of outlier documents with different classification noise filtering approaches. Classification noise filtering, initially suggested by (Brodley & Friedl, 1999), is based on the idea of using a classifier as a tool to detect noisy and outlier instances in the data. In the work of Sluban et al. (2012), the simple classifiers used by Brodley and Friedl (1999) were replaced by new better performing classifiers. To this end, three outlier

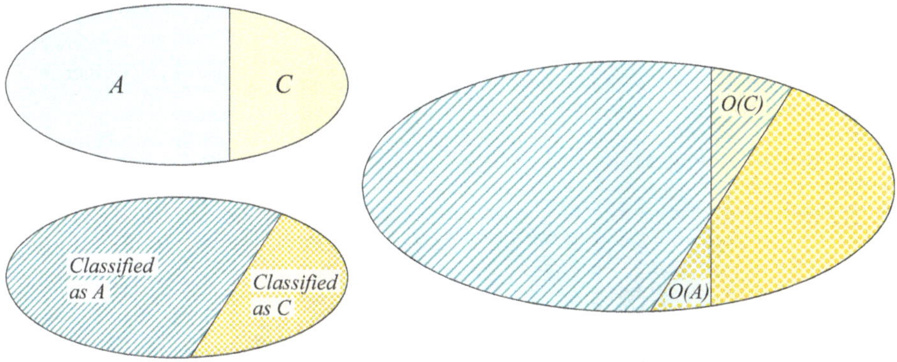

Fig. 6.1 Detecting outliers of a domain pair dataset using document classification.

detection algorithms were implemented using three different classifiers: Naïve Bayes (abbreviated: Bayes), Random Forest (RF) and Support Vector Machine (SVM). These approaches were considered in an ensemble classifier, incorporated in the NoiseRank outlier detection approach (Sluban et al., 2014).

When applying these classifiers, incorrectly classified instances are considered outlier documents. In other words, if an instance of class A is classified in the opposite class C, we consider it an outlier of domain A, and vice versa. We denote the two sets of outlier documents by $O(A)$ and $O(C)$, respectively. Figure 6.1 illustrates this outlier document detection principle.

In addition to the outlier sets obtained by three individual classification algorithms, we also examined the union of these outlier sets and the so-called "Majority" outlier set containing outlier documents that were detected by at least two of the three classification filters. The approach relying on the "Majority" outlier set proved to have the greatest potential for bridging term detection: firstly, because of the best ratio among the proportion of the size of the outlier subset and the proportion of b-terms which are present in that outlier subset, and secondly, because the relative frequency of all the b-terms present in the "Majority" outlier set is higher compared to the entire migraine-magnesium and autism-calcineurin datasets, respectively.

To measure the relevance of the detected outlier documents in terms of their potential to contain bridging terms, we inspected 43 terms known as bridging terms appearing in the migraine-magnesium domain pair and 13 known b-terms in the autism-calcineurin domain pair. We evaluated whether domain outliers obtained by classification algorithms have the potential for bridging term detection, tested on the *migraine-magnesium* (Swanson et al., 2006) and *autism-calcineurin* (Petrič et al., 2009) domain pair datasets, which have lists of confirmed b-terms (see the list of b-terms in Table 4.2 of Chapter 4).

Figure 6.2 shows that very favorable results were obtained for both (migraine-magnesium and autism-calcineurin) datasets. The experimental results show that the sets of detected outlier documents were relatively small (including less than 5% of entire datasets) and that they contained a large majority of bridging terms, which was significantly higher than in same sized random subsets. These results are summarized

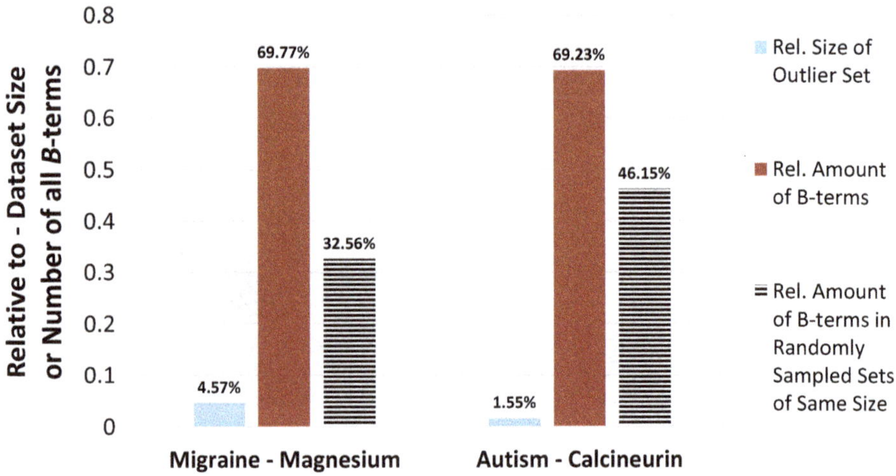

Fig. 6.2 Presence of b-terms in the detected outlier sets of two domain pair datasets.

in Figure 6.2. Therefore, the effort needed for finding cross-domain bridging terms is substantially reduced, as it requires exploring a much smaller subset of documents, where a great majority of b-terms are present and more frequent.

To confirm that these results are not due to chance (do not hold for just any arbitrary subset that has the same size as an outlier set), we have randomly sampled 1,000 subsets for each of the outlier sets (all of them having the same size as their corresponding outlier set) to present the average b-term occurrences in randomly sampled subsets. The results showed that sets of outlier documents contain on average more than 30% more of all b-terms in the migraine-magnesium dataset and more than 20% more of all b-terms in the autism-calcineurin dataset than randomly sampled sets of the same size.

6.3 Document clustering for outlier detection and b-term identification

Petrič et al. (2012a) proposed to search for b-terms in outlier documents, where document clustering was used to find outlier documents, i.e. a specific type of outlier documents (domain outliers) that tend to be more similar to the documents of the opposite domain than to the other documents of their own domain.

Document clustering refers to clustering methods (Kaufman & Rousseeuw, 1990), which are very popular for grouping instances in terms of their similarity, adopted to grouping of text documents (Feldman & Sanger, 2006). In particular, the OntoGen document clustering tool (Fortuna et al., 2006) was used by Petrič et al. (2012a) to find outlier documents.

Let us first briefly describe the OntoGen tool, which was briefly mentioned in Section 3.3.2. OntoGen (Fortuna et al., 2006) is a semi-automatic data-driven interactive text mining tool that assists users in the creative process of creating topic ontologies. It is essentially a text mining tool for grouping documents into related clusters that can be viewed as concepts in an automatically created topic ontology. The underlying methodology is k-means clustering. This technique is particularly popular, as only parameter k needs to be chosen to determine the number of categories into which documents should be clustered. A particularly attractive feature is OntoGen's user-friendly human-computer interface. The 'main window' provides a visualization of the ontology, where each concept is represented by the first three keywords (automatically assigned names of clusters that can be edited manually), while the 'concept hierarchy' window provides a quick outline of all concepts with their position in the concept hierarchy, which can also be manipulated directly.

In the outlier detection approach to LBD, described in this section, each document of the two literatures is represented by a set of words using frequency statistics based on the Bag-of-Words (BoW) text representation, introduced in Section 3.3.1. The BoW vector representation, using the standard TF-IDF (term frequency–inverse document frequency, see Equation 3.1) word weighting scheme (Salton & Buckley, 1988), enables measuring the content similarity of documents. In OntoGen, contents similarity is computed using the cosine distance between the TF-IDF weighted BoW vectors, where a high frequency of co-occurring words in documents indicates high document similarity. In particular, the cosine similarity measure is used to position the documents according to their similarity to the representative document (centroid) of a selected domain. Documents, positioned using the cosine similarity measure, can be visualized in OntoGen with a similarity graph with cosine similarity values that fall within the $[0, 1]$ interval. Value 0 means extreme dissimilarity, where two documents do not share any common words, while value 1 represents the similarity between two semantically identical documents in the BoW representation (see Figure 6.5).

The approach originally proposed by Petrič et al. (2012a) consists of two steps. In the first step, the OntoGen clustering algorithm is applied to cluster the merged document set $A \cup C$, consisting of documents from both domains A and C. The result of unsupervised clustering is a set of two document clusters: C' (i.e. a set of documents from $A \cup C$ classified as C), and A' (i.e. a set of documents from $A \cup C$ classified as A). Then, in the second step, for each of the clusters, a supervised clustering approach is applied taking into account documents' original domains C and A. As a result, a two-level tree hierarchy of clusters is generated, as illustrated in Figures 6.3 and 6.4.

Conceptually, clustering can be used to create two document clusters, C' and A' (where $A' \cup C' = AC$) in an unsupervised manner, using OntoGen's 2-means clustering algorithm (see Figure 6.3). Cluster C' (labeled "Classified as C" in Figure 6.3) consists mainly of documents from C, but may also contain some documents from A. Similarly, cluster A' (labeled "Classified as A" in Figure 6.3) consists mainly of documents from A, but may also contain some documents from C. The approach, where OntoGen is used for document clustering, is illustrated in Figure 6.4.

The experimental results obtained in the gold standard migraine-magnesium domain, as well as in the autism-calcineurin domain pair, confirm the hypothesis that most bridging terms occur in outlier documents and that by considering only outlier documents the search space for b-term identification can be largely reduced. To this end, the user can explore the document similarity graph (schematically presented in

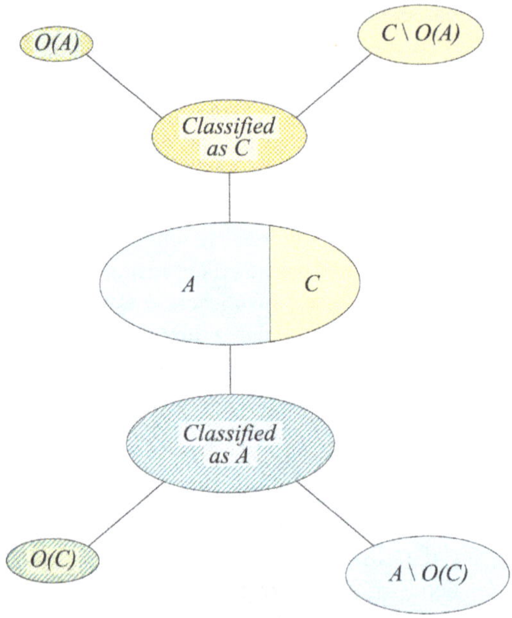

Fig. 6.3 Documents from literatures C and A, clustered according to the OntoGen's two-step approach, first using unsupervised and then supervised clustering to obtain outlier documents $O(C)$ and $O(A)$ of literatures C and A, respectively.

Figure 6.5) to identify the most irregular documents of their own domain, and start searching for b-terms from outlier documents in these subclusters O(C) and O(A). This can significantly reduce the search space for finding b-term candidates.

6.4 Relating outlier detection to CrossBee heuristics

The results of Sluban et al. (2012) and of Petrič et al. (2012a) confirm the hypothesis that most bridging terms occur in outlier documents and that by considering only outlier documents the search space for b-term candidates can be largely reduced. This has two important consequences. The first consequence of reducing the search space is improved efficiency in searching for b-terms, which is manifested in reduced time required for the task. The second, possibly more important consequence, is the ability to work with originally larger document bodies, since the resulting corpora of outlier documents are usually much smaller than the original ones.

Given this insight, outlier document detection approaches, described in Section 6.2 and Section 6.3, have inspired the development of outlier-based heuristics of the CrossBee b-term detection engine, described in Section 5.2.

As mentioned in Section 5.2.2, six heuristics (outFreqRelRF, outFreqRelSVM, outFreqRelCS, outFreqSum, tfidfDomnSum, freqRatio) are used in the CrossBee ensemble

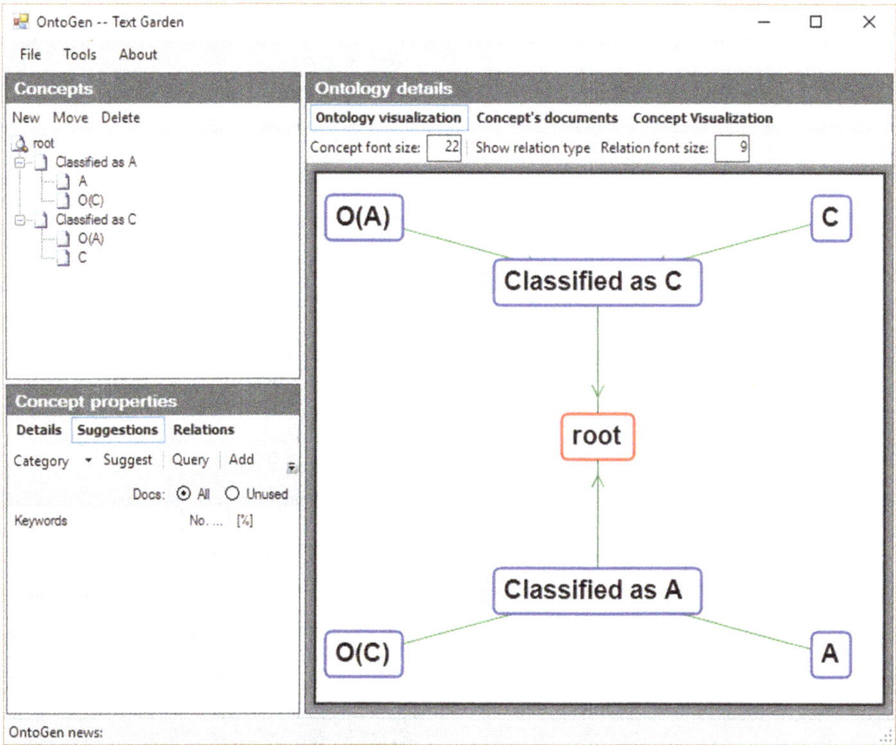

Fig. 6.4 Target domain documents from literatures C and A, clustered according to the OntoGen's two-step approach, first using unsupervised and then supervised clustering to obtain outlier documents $O(C)$ and $O(A)$ of literatures C and A, respectively.

heuristic. The outlier-based heuristics proved to be very effective. Note that four of these (outFreqRelRF, outFreqRelSVM, outFreqRelCS, outFreqSum) are based on term frequencies in outlier documents, three of which were inspired by the classification-based approach (outFreqRelRF, outFreqRelSVM, outFreqRelCS) and one by the OntoGen clustering approach (outFreqSum) to outlier document detection.

6.5 Methodology combining outlier detection and CrossBee document exploration

Compared to the outlier document detection approach using NoiseRank (outlined in Section 6.2) or OntoGen (outlined in Section 6.3), a combined approach is illustrated in this section. To illustrate the combined outlier-based LBD methodology, we use the workflow approach used in the web-based text mining platform TextFlows[1] (Perovšek

[1] http://textflows.org

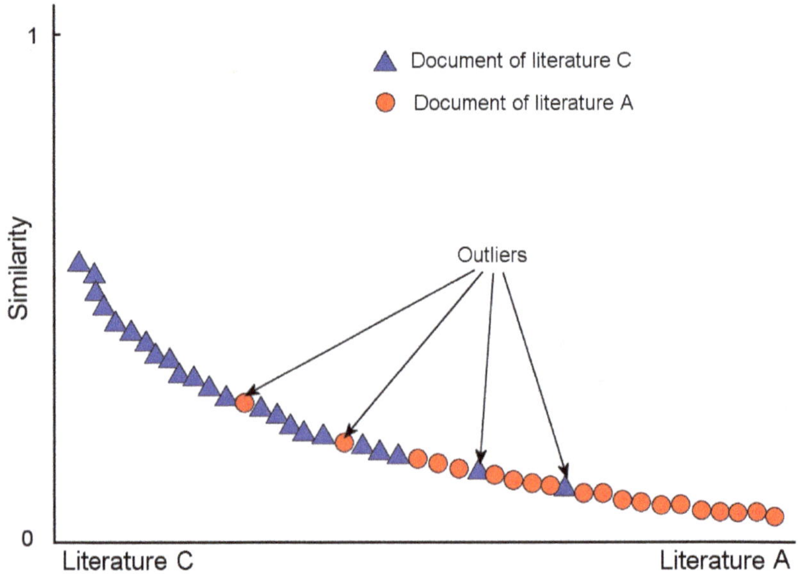

Fig. 6.5 Graph representing instances (documents) of literature C (blue triangles) and instances (documents) of literature A (red circles) according to their content similarity to a prototypical document c of literature C. In this similarity graph, outliers of literature A are positioned closer to the typical representatives of literature C than to the central documents of literature A (i.e. the outliers of domain A most similar to the centroid c of domain C are the two red circles positioned among the blue triangles; these two red circles represent $O(A)$ positioned in the left part of the similarity graph).

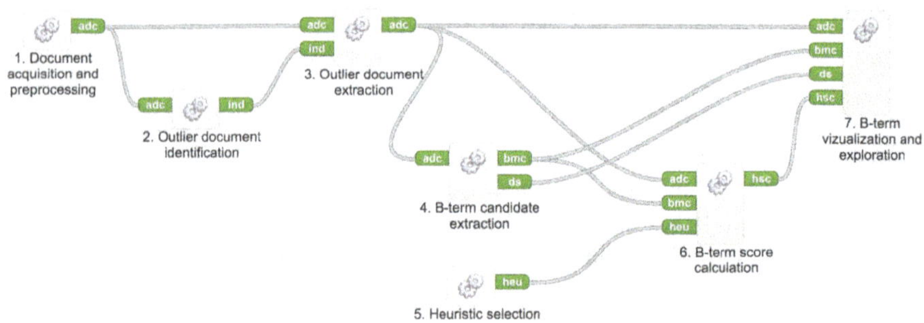

Fig. 6.6 A top-level workflow of the proposed methodology in TextFlows (Perovšek et al., 2016). The acronyms of component outputs and inputs denote the connector's data type.

et al., 2015), which was developed to construct and execute advanced text mining pipelines. The workflow shown in Figure 6.6 consists of seven steps that define the main subprocesses. The connections between subprocesses represent the flow of documents from one subprocess to another. In summary, steps 1–3 represent the outlier detection part, and steps 4–7 represent the cross-domain exploration part.

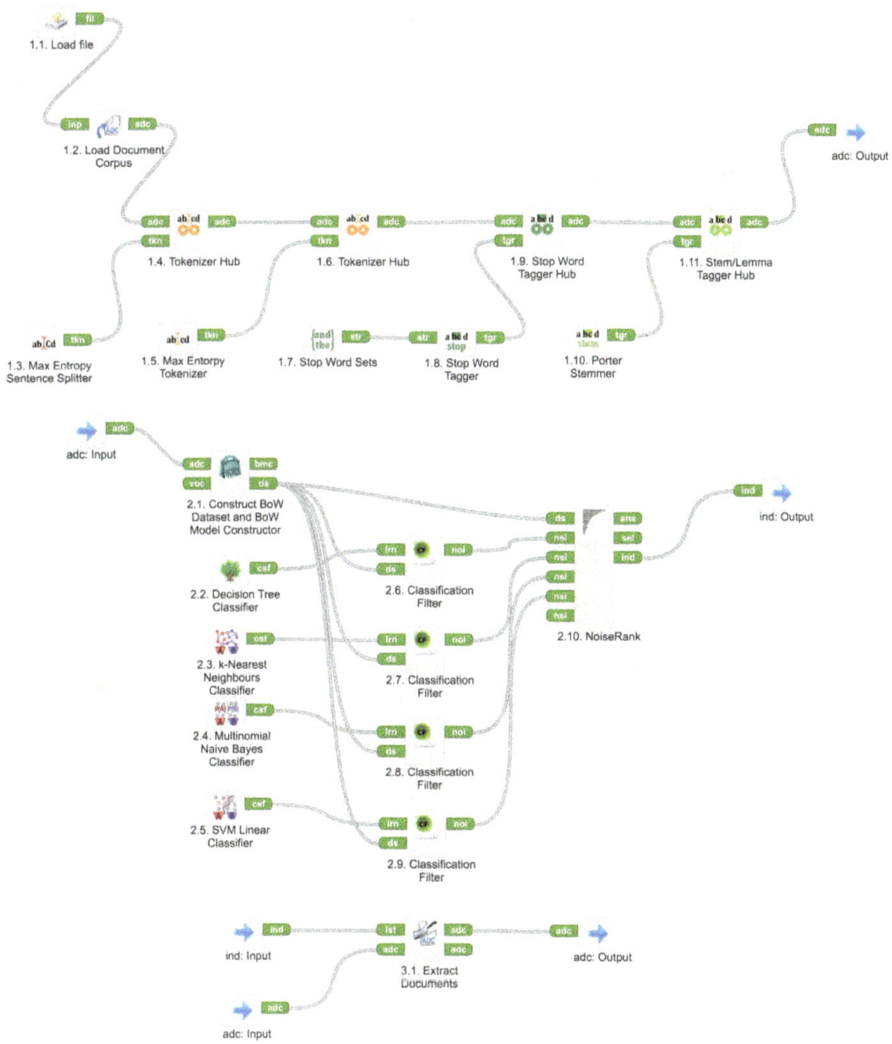

Fig. 6.7 Detailed workflows for the first three sub-processes that are used to identify and extract outlier documents.

- In the first three steps of the workflow shown in Figure 6.6, the outlier documents are identified and extracted using the NoiseRank algorithm (Sluban et al., 2014). The goal of the first phase is to extract a set of outlier documents from the entire corpus of input documents, thus reducing the size of the input set of documents, which enables the second phase of the LBD process to become more focused, efficient, and effective. Figure 6.7 shows the details of the first three steps of the workflow.

- In the last four steps of the workflow shown in Figure 6.6, the CrossBee components (Juršič et al., 2012a) are executed in expert-guided b-term analysis. Here, the goal is to further prepare the input documents for b-term visualization and exploration, where the role of the domain expert is crucial. Figure 6.8 shows details of the last four steps of the workflow.

Figure 6.7 presents detailed workflows for the first three subprocesses: 1. *Document acquisition and preprocessing*, 2. *Outlier document identification*, and 3. *Outlier document extraction*.

1. **Document acquisition and preprocessing.** The task of the first subprocess is to read the input set of documents from a file and convert the text documents into a well-structured data representation, composed of a set of relevant features for further processing (Feldman & Sanger, 2006). Each document is represented by a BoW vector of numerical values (Sebastiani, 2002), one for each feature of the selected representational model. The illustrated text preprocessing techniques are based on the standard text mining concepts, including Tokenization, Stop words tagging, and Stem/Lemma tagging, which are all present in the workflow of the first subprocess. Note that the preprocessing techniques used do not include more complex text mining techniques such as Part-Of-Speech tagging, Entity detection, and Relation detection (Feldman & Sanger, 2006), nor the contemporary text embedding techniques outlined in Section 3.5.
2. **Identification of outlier documents.** The purpose of the second subprocess is to identify outlier documents from the initial document corpus, using the NoiseRank approach (Sluban et al., 2014), outlined in Section 6.2. Here, classifiers are used to detect atypical documents in categorized document corpora, which can be considered to be outliers of their own document category. In the workflow shown in Figure 6.7 we used four classifiers: 2.2. *Decision Tree classifier*, 2.3. *k-Nearest Neighbors classifier*, 2.4. *Multinomial Naïve Bayes classifier*, and 2.5. *SVM Linear classifier* (Weeber et al., 2001). Outlier documents identified by these classifiers are used as input to 1.10. *NoiseRank*, which supports domain experts in identifying noisy, outlier or erroneous data instances (Sluban et al., 2014). Users can select the noise detection algorithms to be used in the ensemble-based noise detection process. The NoiseRank methodology workflow returns a visual representation of a list of potential outlier documents, ranked according to the decreasing number of noise detection algorithms, which identified a document as an outlier. So, in addition, the users can obtain a visual representation of top-ranked outlier documents, allows the user's manual intervention (inclusion or exclusion from the list of outliers).
3. **Extraction of outlier documents.** This process extracts the identified outlier documents from the initial document corpus.

Figure 6.8 presents the workflows for subprocesses 4. *b-term candidate extraction*, 5. *Heuristics selection*, 6. *b-term score calculation*, and 7. *b-term visualization and exploration*.

4. **Extraction of *b*-term candidates.** Component 4.1 constructs the BoW representation of the extracted outlier documents.

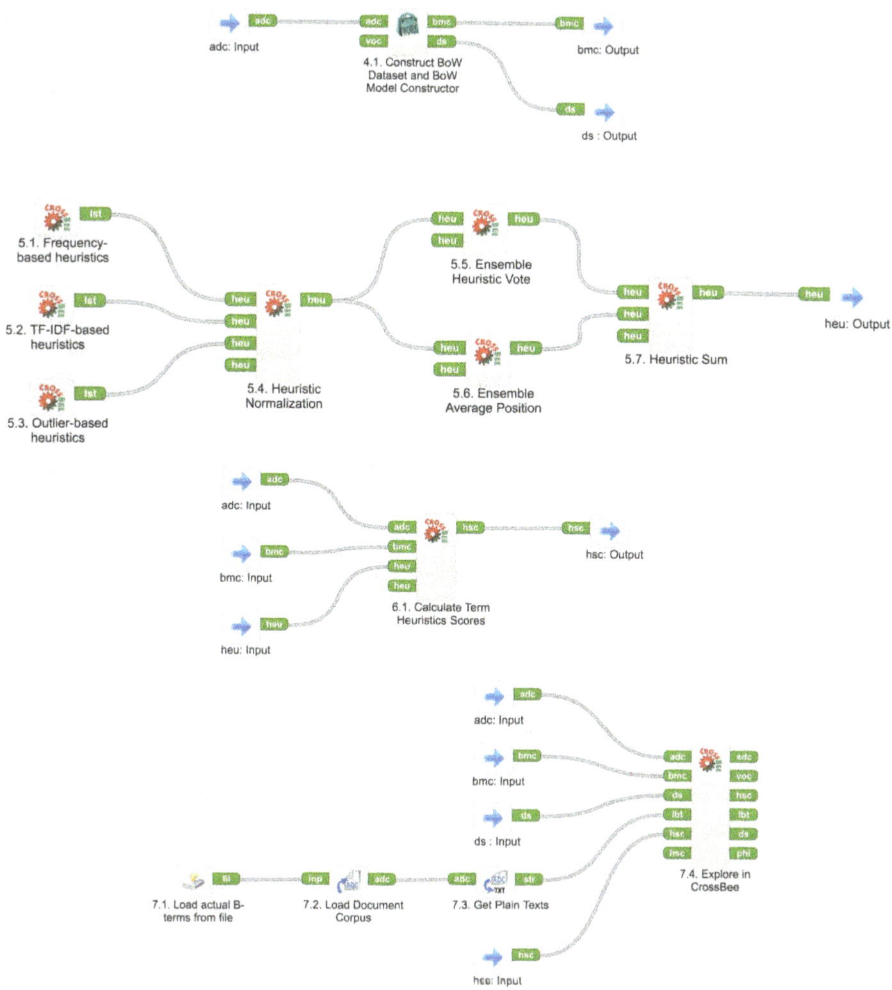

Fig. 6.8 Workflows for the last four sub-processes implementing the individual CrossBee components.

5. Selection of heuristics. In this subprocess, heuristics used to evaluate the b-term potential are selected (Juršič et al., 2012b) whose results are combined to obtain the score. Using Equation 5.2 presented in Section 5.2.2.

6. b-term score calculation. This sub-process consists of a single component labeled 6.1. *Calculate Term Heuristic Scores*. This component takes as an input several heuristics specifications and performs the actual calculations of the heuristics on every b-term.

7. *b*-term visualization and exploration. The most important component of this subprocess is 7.4. *Explore in CrossBee*, which exports the final ranking results and the annotated document corpus to the web application CrossBee. This component, illustrated in Figure 5.6 presented in Section 5.2.3, enables effective manual exploration of potential b-terms and the respective documents.

6.6 Application: Finding connections between Alzheimer's disease and gut microbiota

In the practical experiment using the combined methodology, we demonstrated the utility of outlier detection to find links between two distinctive domain literatures: Alzheimer's disease and human gut microbiota.

In the experiment, we downloaded two sets of documents from PubMed: 83,322 papers obtained from the query "Alzheimer" and 73,960 papers obtained from the query "(gut OR intestinal) AND (microbiota OR bacteria)". We used titles and abstracts based on previous experimental evidence described by Petrič et al. (2006). Due to the functional limitations of the CrossBee tool, these sets were reduced by eliminating documents older than two years (i.e. we included years 2014 and 2015) and documents with incomplete title or abstract. As a result, we obtained 8,934 "Alzheimer" papers and 8,937 "gut microbiota" papers. As a proof of concept, following the steps of the workflow illustrated in Figure 6.7), we first extracted a random sample of 1,971 papers from the entire set of 17,863 papers (11%) and obtained 248 outlier documents for further processing.

For efficiency reasons, instead of using NoiseRank, we decided to use OntoGen for outlier document detection. From the joint set of 17,863 papers we constructed a two-level document hierarchy with OntoGen, following the approach to identify outlier documents, proposed by Petrič et al. (2012a) and outlined in Section 6.3. At the first level, after transforming the documents into a feature vector format, the documents were clustered according to the cosine similarity into two distinct document clusters. At the second level, as shown in Figure 6.3, each of the two clusters was further separated according to the document search origin ("Alzheimer" or "gut microbiota") into two sub-clusters.

Based on the constructed cluster hierarchy from Figure 6.3 we obtained 582 outlier documents (428 from "Alzheimer" and 154 from "gut microbiota"). We further explored these documents in the CrossBee tool. By processing the documents from the two separate domains (literatures) of interest, CrossBee extracted 4,723 terms as potential terms that connect the two domains. The terms in CrossBee were ranked in terms of their estimated bridging term potential (Juršič et al., 2012a), tending to push more interesting terms to the top of the list. Although the list of potential bridging terms is ordered according to their term potential, browsing and analyzing the terms from the list still presents a substantial burden for the domain expert and supportive team.

To test the methodology, the domain expert manually created a list of 289 potentially interesting terms, which were manually extracted from 42 articles preselected by the expert. General terms (such as dementia, aging, neurodegeneration, inflammation, microbiome, probiotics, etc.) and specific molecular factors and pathways (such as the names of transcription factors, trophic factors, misfolded or aggregated proteins, mechanisms of oxidative stress and lipid synthesis, etc.) were manually identified in the title, abstracts, and keywords of these publications. 55 terms from this list appeared among the 4,723 terms extracted by CrossBee.

Several terms suggested by CrossBee and identified by the expert as promising were further explored. For example, potential terms included *cox* (ranked 80 in CrossBee), *mucosa* (122), *beta amyloid* (749), *nitric oxide synthase* (1,153), and *phenylalanine* (2,080). In subsequent evaluation and discussion, the relevant publications for each b-term candidate were reviewed and searched for potential clues that triggered further investigation. At the time of publication in 2016, the most promising identified b-term was 'nitric oxide synthase' (i.e. NOS) (Katusic & Austin, 2014), which can be interpreted in terms of the contribution of the microbiome in NOS-mediated inflammation at the level of the gut (Baruch et al., 2015; Derkinderen et al., 2011; Xiao et al., 2014), which likely disregulates brain-gut communication. Under pathological conditions, such as inflammation or in the presence of environmental stressors or aging, abnormal iNOS function leads to oxidative effects and neurodegeneration. In particular, in the gut, iNOS leads to intestinal barrier damage (Grishin et al., 2016) and in the brain to protein nitrosylation and cell death with neurological consequences such as dementia, Alzheimer's disease, or Parkinson's disease (Grishin et al., 2016; Hess et al., 2005). Even when NO is produced locally, its effects can diffuse systemically and influence disease progression (Gan et al., 2015; Mancuso & Santangelo, 2014; Rannikko et al., 2015). Bioactive nutrients could modulate individual microbiome responses by limiting inflammation and free radical synthesis, including NO (Jeong et al., 2015). Further studies are needed to elucidate the consequences of pathological NO signaling in various tissues to explore the clinical utility of NO /iNOS-targeted therapeutic strategies (Broom et al., 2011).

In the experiments, a biomedical domain expert had a crucial role in efficient document exploration when investigating the bridging terms proposed by CrossBee. In the experiment, where the term *nitric oxide synthase* (Katusic & Austin, 2014) was investigated as a promising new bridging term from the set of articles used as input to CrossBee, the expert carefully reviewed the four articles containing this term: three from the *Alzheimer* domain (Gan et al., 2015; Mancuso & Santangelo, 2014; Rannikko et al., 2015) and one from the *gut microbiota* domain (Xiao et al., 2014). Moreover, the combined query "Alzheimer gut nitric oxide" in PubMed revealed that there were no articles matching this query at the time of the experiment in 2016, which was particularly interesting, suggesting that this association has not been investigated in the available literature by the date of our experiment.

In the first article by Ma et al. (2018), endothelial nitric oxide synthase (eNOS) is mentioned as part of the neurovascular improvements observed in mice following a ketogenic diet. Specifically, the ketogenic diet has been associated to increased expression of eNOS proteins. eNOS is involved in the production of nitric oxide, which plays a critical role in vascular functions such as blood flow regulation and may influence neurodegeneration risk in Alzheimer's disease.

The second article by Bonfili et al. (2021) mentions enhanced cerebral expression of hypoxia-inducible factor-1α (HIF-1α) by chronic supplementation with SLAB51 in mice. Although endothelial nitric oxide synthase (eNOS) is not directly mentioned, the link between HIF-1α and nitric oxide levels is suggested by the decrease in prolyl hydroxylase 2 (PHD2), an oxygen-dependent regulator of HIF-1α degradation, and the decrease in the expression of inducible nitric oxide synthase (iNOS) in the brain. iNOS produces large amounts of nitric oxide under inflammatory conditions, which can influence neuroinflammation as a factor in the progression of Alzheimer's disease.

The third article by Salminen (2023) deals with the activation of the aryl hydrocarbon receptor (AhR) in Alzheimer's disease and its consequences for cerebrovascular integrity. In particular, it mentions that AhR signaling inactivates endothelial nitric oxide synthase (eNOS), thus impairing nitric oxide production and vasodilation. This inactivation contributes to the pathology of Alzheimer's disease by promoting vascular dysfunction, including inflammation and oxidative stress.

From these three articles that present related studies, it is clear that nitric oxide synthase, particularly eNOS, plays an important role in the integrity of neurovascular function and the pathophysiology of Alzheimer's disease, which is influenced by changes in diet or microbiota.

Note that later, at the time of writing this monograph, using the same PubMed query, we found 3 articles in the PubMed literature that also contained the term *nitric oxide synthase* that were published well after our original publication (Cestnik et al., 2016), demonstrating the potential of the outlined methodology for new bisociative discoveries.

6.7 Discussion

In summary, our study showed that the LBD process supported by combining outlier detection with CrossBee was able to identify novel common molecular targets that can be used as multipurpose drugs capable of blocking multiple processes such as oxidative and inflammatory effects with high neuroprotective relevance to peripheral and central neurons. Note that the combined query "Alzheimer gut nitric oxide" in PubMed revealed that there were no articles matching this query, suggesting that at the date of this study this relationship had not been previously investigated in the PubMed literature.

The presented methodology integrates the process of exploring outlier documents and the literature-based discovery process. The methodology was first executed by combining two text mining tools OntoGen (Fortuna et al., 2006) and CrossBee (Juršič et al., 2012a). Apart from constructing the document clustering from the input set of documents and finding the keywords describing the two document classes, OntoGen can be used to narrow down the search for bridging terms in CrossBee by identifying the set of outlier documents. In such a way, the search in CrossBee can be more focused, efficient, and effective. This opens the door to many new applications in which the size of the search space is a severe limiting factor.

We have also presented the methodology as a workflow in the TextFlows platform (Perovšek et al., 2016), to demonstrate and explain the details of the procedural steps required by the methodology. In the illustrative workflow, OntoGen-based outlier detection was replaced by a classification-based approach using NoiseRank. While the two approaches share the same philosophy, they are not identical because they use different methods to detect outlier documents. The CrossBee part of the methodology corresponds to subprocesses 4–7 in the TextFlows workflow in Figure 6.6. However, the outlier detection part (subprocesses 1–3 in Figure 6.6) differs between the two approaches. OntoGen is efficient and can process several thousands of documents in a couple of minutes, however, it might take the classifier-based NoiseRank approach several hours to process the same workload. So, for efficiency reasons, the outlier detection part is in practice best performed by OntoGen.

6.8 Tutorial in Python

The content presented in this chapter is supplemented by a practical example whose source code is hosted on GitHub. We recommend the reader to access the tutorial **04_mini_ontogen.ipynb** in the Github repository **https://github.com/lbd-book/code/** to further explore the concepts discussed here. Before running the Python notebook, we suggest that you first read Chapter 4 and follow the instructions to install Python and the required packages from the **requirements.txt** file in the repository.

6.8.1 04_mini_ontogen.ipynb

Outlier-based LBD methods for closed discovery efficiently narrow down the search for cross-domain bridging terms to outlier documents, thus simplifying the exploration of bridging terms by domain experts. The workflow illustrates the experiments with the OntoGen tool (Cestnik et al., 2017), which is used in b-term discovery by detecting outlier documents in the Autism–Calcineurin domain pair (Petrič et al., 2012b; Sluban et al., 2012).

The process starts with cleaning and structuring the text data, followed by feature extraction using a BoW model and a TF-IDF matrix, both of which are refined by filtering out less relevant terms. The reduction of dimensionality with principal component analysis enables two-dimensional visualization of document clusters determined by k-means clustering. The combinations of domains and clusters are analyzed to detect overlaps and identify outlier groups (i.e. Autism [Cluster #1] and Calcineurin [Cluster #0] with 296 and 73 documents, respectively). The workflow concludes with the storage of these outliers, which are saved in a file for further processing and experimentation. The illustrated method is implemented in the `04_mini_ontogen.ipynb` notebook, available in the repository of this book: `https://github.com/lbd-book/code/`. Note that by setting the variable `CONTEXT_SWITCH` in the notebook to 2, the corresponding workflow illustrates the experiments with OntoGen in the Alzheimer–gut microbiota domain pair.

References

Aggarwal, C. (2013). *Outlier Analysis*. Springer.

Baruch, K., Kertser, A., Porat, Z., & Schwartz, M. (2015). Cerebral nitric oxide represses choroid plexus NFκB-dependent gateway activity for leukocyte trafficking. *The EMBO Journal*, *34*(13), 1816–1828.

Bonfili, L., Gong, C., Lombardi, F., Cifone, M. G., & Eleuteri, A. M. (2021). Strategic modification of gut microbiota through oral bacteriotherapy influences hypoxia inducible factor-1α: Therapeutic implication in alzheimer's disease. *International journal of molecular sciences*, *1*(23), 357.

Brodley, C. E., & Friedl, M. A. (1999). Identifying mislabeled training data. *Journal of Artificial Intelligence Research*, *11*, 131–167.

Broom, L., Marinova-Mutafchieva, L., Sadeghianand, M., Davis, J. B., Medhurst, A. D., & Dexter, D. T. (2011). Neuroprotection by the selective iNOS inhibitor GW274150 in a model of Parkinson disease. *Free Radical Biology and Medicine*, *50*(51), 633–640.

Cestnik, B., Fabbretti, E., Gubiani, D., Lavrač, N., & Urbančič, T. (2016). Exploring outlier documents to investigate potential links between gut microbiota and Alzheimer's disease. *International Work-Conference on Bioinformatics and Biomedical Engineering, IWBBIO 2016*, 475–486.

Cestnik, B., Fabbretti, E., Gubiani, D., Urbančič, T., & Lavrač, N. (2017). Reducing the search space in literature-based discovery by exploring outlier documents: A case study in finding links between gut microbiome and Alzheimer's disease. *Genomics and Computational Biology*, *3*(3), e58.

Derkinderen, P., Rouaud, T., Lebouvier, T., Bruley des Varannes, S., Neunlist, M., & De Giorgio, R. (2011). Parkinson disease: The enteric nervous system spills its guts. *Neurology*, *77*(19), 1761–1767.

Feldman, R., & Sanger, J. (2006). *The text mining handbook: Advanced approaches in analyzing unstructured data*. Cambridge University Press.

Fortuna, B., Grobelnik, M., & Mladenić, D. (2006). Semi-automatic data-driven ontology construction system. *Proceedings of the 9th International Multiconference Information Society*, 223–226.

Gan, P., Zhang, L., Chen, Y., Zhang, Y., Zhang, F., Zhou, X., Zhang, X., Gao, B., Zhen, X., Zhang, J., & Zheng, L. T. (2015). Anti-inflammatory effects of glaucocalyxin B in microglia cells. *Journal of Pharmacological Sciences*, *128*(1), 35–46.

Grishin, A., Bowling, J., Bell, B., Wang, J., & Ford, H. R. (2016). Roles of nitric oxide and intestinal microbiota in the pathogenesis of necrotizing enterocolitis. *Journal of Pediatric Surgery*, *51*(1), 13–17.

Hess, D. T., Matsumoto, A., Kim, S. O., Marshall, H. E., & Stamler, J. S. (2005). Protein S-nitrosylation: Purview and parameters. *Nature Reviews Molecular Cell Biology*, *6*, 150–166.

Jeong, J. J., Woo, J. Y., Ahn, Y. T., Shim, J. H., Huh, C. S., Im, S. H., Han, M. J., & Kim, D. (2015). The probiotic mixture IRT5 ameliorates age-dependent colitis in rats. *International Immunopharmacology*, *26*(21), 416–422.

Juršič, M., Cestnik, B., Urbančič, T., & Lavrač, N. (2012a). Cross-domain literature mining: Finding bridging concepts with CrossBee. *Proceedings of the 3rd International Conference on Computational Creativity*, 33–40.

Juršič, M., Cestnik, B., Urbančič, T., & Lavrač, N. (2012b). Bisociative literature mining by ensemble heuristics. In M. R. Berthold (Ed.), *Bisociative knowledge discovery* (pp. 338–358). Springer.

Katusic, Z. S., & Austin, S. A. (2014). Endothelial nitric oxide: Protector of a healthy mind. *European Heart Journal*, *35*(14), 888–894.

Kaufman, L., & Rousseeuw, P. J. (1990). *Finding groups in data: An introduction to cluster analysis*. John Wiley & Sons.

Ma, D., Wang, A. C., Parikh, I., Green, S. J., Hoffman, J. D., Chlipala, G., Murphy, M. P., Sokola, B. S., Bauer, B., Hartz, A. M. S., & Lin, A. L. (2018). Ketogenic diet enhances neurovascular function with altered gut microbiome in young healthy mice. *Scientific reports*, *1*(8).

Mancuso, C., & Santangelo, R. (2014). Ferulic acid: Pharmacological and toxicological aspects. *Food and Chemical Toxicology*, *65*, 185–195.

Moore, D. S., McCabe, G. P., & Craig, B. A. (2009). *Introduction to the Practice of Statistics*. WH Freeman New York.

Perovšek, M., Kranjc, J., Erjavec, T., Cestnik, B., & Lavrač, N. (2016). Textflows: A visual programming platform for text mining and natural language processing. *Special Issue on Knowledge-based Software Engineering, Science of Computer Programming*, *121*, 128–152.

Perovšek, M., Podpečan, V., Kranjc, J., Erjavec, T., Pollak, S., Do Thi, N. Q., Liu, X., Smith, C., Cavazza, M., & Lavrač, N. (2015). Text mining platform for NLP workflow design, replication and reuse. *Proceedings of IJCAI Workshop on Replicability and Reusability in Natural Language Processing: From Data to Software Sharing*.

Petrič, I., Cestnik, B., Lavrač, N., & Urbančič, T. (2012a). Outlier detection in cross-context link discovery for creative literature mining. *The Computer Journal*, *55*(1), 47–61.

Petrič, I., Cestnik, B., Lavrač, N., & Urbančič, T. (2012b). Bisociative knowledge discovery by literature outlier detection. In M. R. Berthold (Ed.), *Bisociative knowledge discovery: An introduction to concept, algorithms, tools, and applications* (pp. 313–324). Springer.

Petrič, I., Urbančič, T., & Cestnik, B. (2006). Comparison of ontologies built on titles, abstracts and entire texts of articles. *Proceedings of the 9th International Multi-conference Information Society*, 227–230.

Petrič, I., Urbančič, T., Cestnik, B., & Macedoni-Lukšič, M. (2009). Literature mining method RaJoLink for uncovering relations between biomedical concepts. *Journal of Biomedical Informatics*, *42*(2), 219–227.

Rannikko, E. H., Weber, S. S., & Kahle, P. J. (2015). Exogenous α-synuclein induces toll-like receptor 4 dependent inflammatory responses in astrocytes. *BMC Neuroscience*, 16–57.

Salminen, A. (2023). Activation of aryl hydrocarbon receptor (ahr) in alzheimer's disease: Role of tryptophan metabolites generated by gut host-microbiota. *Journal of molecular medicine (Berlin, Germany)*, *35*(101), 201–222.

Salton, G., & Buckley, C. (1988). Term-weighting approaches in automatic text retrieval. *Information Processing and Management*, *24*(5), 513–523.

Sebastiani, F. (2002). Machine learning in automated text categorization. *ACM Computing Surveys*, *34*(1), 1–47.

Sluban, B., Juršič, M., Cestnik, B., & Lavrač, N. (2012). Exploring the power of outliers for cross-domain literature mining. In M. R. Berthold (Ed.), *Bisociative Knowledge Discovery* (pp. 325–337). Springer.

Sluban, B., Gamberger, D., & Lavrač, N. (2014). Ensemble-based noise detection: Noise ranking and visual performance evaluation. *Data Mining and Knowledge Discovery*, *28*(2), 265–303.

Swanson, D. R., Smalheiser, N. R., & Torvik, V. I. (2006). Ranking indirect connections in literature-based discovery: The role of Medical Subject Headings (MeSH). *Journal of the American Society for Information Science and Technology*, *57*(11), 1427–1439.

Weeber, M., Klein, H., de Jong-van den Berg, L., Vos, R., et al. (2001). Using concepts in literature-based discovery: Simulating Swanson's Raynaud–fish oil and migraine–magnesium discoveries. *Journal of the American Society for Information Science and Technology*, *52*(7), 548–557.

Xiao, J., Li, S., Sui, Y., Wu, Q., Li, X., Xie, B., Zhang, M., & Sun, Z. (2014). Lactobacillus casei-01 facilitates the ameliorative effects of proanthocyanidins extracted from lotus seedpod on learning and memory impairment in scopolamine-induced amnesia mice. *PLoS One*, *9*(11), e112773.

Chapter 7
Semantic and Outlier-based Open Discovery

In contrast to the closed LBD setting addressed in Chapters 5 and 6, open LBD adds another dimension to the discovery process by allowing that only domain C is predetermined, thus facilitating the search for multiple potential candidates for the domain A via intermediate domains B. To set the ground, the distinction between closed and open LBD is explained in Section 7.1. The first approach combining open and closed discovery is a concept-based approach presented in Section 7.2. The approach is illustrated by replicating Raynaud's syndrome–Fish oil discovery through concept mapping, followed by the presentation of a semantic filtering approach to open discovery. Another approach, combining the text- and outlier-based open LBD, is the RaJoLink approach, presented in Section 7.3, which is illustrated in a concrete application in the autism-calcineurin domain pair. The chapter concludes with Section 7.4, which includes Python tutorials that allow experiment replicability and code reuse.

7.1 Comparison of open and closed LBD

Closed LBD, first established by Swanson's ABC model (Swanson, 1986b), works within a rather narrow framework by taking a pair of domains C and A as provided by the researcher. This methodology focuses on identifying novel connections b within these predetermined boundaries and aims to uncover hidden associations between these fixed domains. While often effective in practice, this approach limits the scope of possible discoveries to the ranges initially selected by the user.

In contrast, open LBD enables a more dynamic exploration of the knowledge landscape. Instead of having to pre-select two initial domains of interest, a researcher has to determine just one, counting on the open LBD method to provide promising candidates for the second domain, through intermediate areas as paths to new knowledge. This proactive strategy strives to increase the potential for serendipitous discovery by leveraging associations between different scientific fields to uncover previously unnoticed connections.

N. Lavrač et al., *Bisociative Literature-Based Discovery*,
https://doi.org/10.1007/978-3-031-96863-1_7

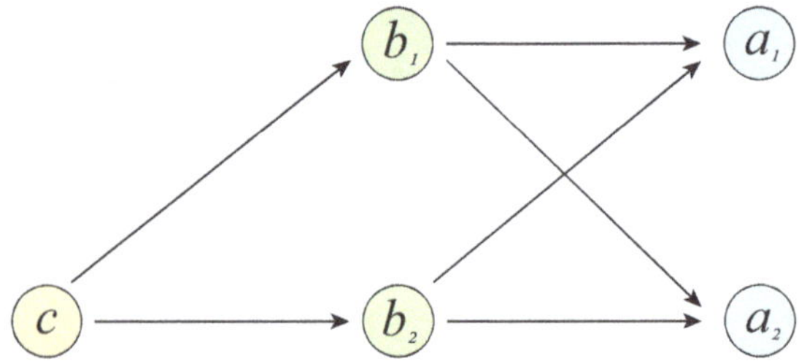

Fig. 7.1 Only the starting concept c is given in open discovery; the goal is to identify target concepts $a_1, a_2, \ldots, a_m$ through intermediate concepts $b_1, b_2, \ldots, b_n$.

In summary, unlike closed LBD that operates within fixed, user-defined domains C and A, open LBD starts with a single known domain C and explores associated domains (candidates for A) via intermediate concepts b from intermediate domains B, thereby expanding the scope of discovery, as illustrated in Figure 7.1. This fundamental difference emphasizes the potential of open LBD to uncover unexpected associations and broader insights across different domains of knowledge. As such, the open discovery process is thus characterized by the *generation* of hypotheses, while the closed discovery process is characterized by the *elaboration* of hypotheses.

In practice, open LBD can be used to help researchers by directing their focus from their own domain C to other seemingly unrelated problem areas. Considering the vastness of the scientific landscape, manually searching through domains to find a promising area of interest A is like searching for a needle in a haystack. For instance, an experienced researcher studying migraines might want to explore beyond the already established relationships (such as magnesium, discovered by Swanson (1986b)), the researcher is eager to explore in other directions. Suppose that an open LBD system suggests a link between migraines (domain C) and seasonal variations (a candidate for domain A). This would encourage the researcher to investigate whether migraine attacks vary with the seasons or occur in response to seasonal changes, an aspect previously not considered relevant. Intrigued by this suggestion, the researcher might embark on a new research path to explore the potential connections between migraine and seasonal variations. This could lead to groundbreaking findings that certain seasonal changes significantly reduce the frequency and severity of migraine attacks.

The next sections present selected LBD approaches. The first approach, which combines open and closed discovery, is concept-based LBD followed by semantic filtering, presented in Section 7.2. Another approach, combining the text- and outlier-based open LBD, is the RaJoLink approach, presented in Section 7.3, which is illustrated with a concrete application in the autism-calcineurin domain pair. Other related open LBD approaches from the literature are described in Chapter 2.

7.2 Concept-based approach to open discovery

Swanson's approach utilized natural language text representation of papers. To address the complexity and intricacies of natural language, various NLP techniques have been utilized in literature-based discovery, some of which were outlined in Section 3.3 and Section 5.1.

Early replicators of Swanson's work used a stop-words list (a list of uninteresting terms) approach to discriminate between interesting and uninteresting n-grams. They excluded general categories such as numerals, prepositions, or adverbs, as well as overly broad terms such as human, disease, or dosage. If any of the stop words appeared, the entire n-gram was removed. For example, the current version of the ARROWSMITH system uses a stop-word list that includes more than 8,000 terms.[1]

Weeber et al. (2001) took a step beyond lexical analysis and applied the Unified Medical Language System (UMLS) Metathesaurus to map free text from the titles and abstracts of MEDLINE papers to UMLS concepts. At that time, the UMLS Metathesaurus contained about 730,000 concepts derived from more than 90 source vocabularies and thesauri. Once the raw text is mapped to the knowledge concepts, one can proceed with either the open or closed literature-based discovery.

This section demonstrates the replication of Swanson's Raynaud's disease-fish oil discovery by incorporating UMLS background knowledge, followed by a brief presentation of the semantic filtering method to reduce the concept space.

7.2.1 Replicating Raynaud's disease–fish oil open discovery through concept mapping

In his pioneering work, the link between fish oil and Raynaud's disease was discovered by chance when manually searching the literature (Swanson, 1986a, 1986b). This section does not attempt to exactly replicate Swanson's results but rather simulates his actual discovery of the link between fish oil and Raynaud's disease, which implies that we do not aim at a complete description of all the intermediate links that were reported by Swanson.

We retrieved 3,058 bibliographic records from PubMed (as of May 17, 2024) using the Entrez query

```
raynaud*[TIAB] AND 1900/01/31:1985/12/31 [PDAT] AND medline [SB]
↪ AND english [LA]
```

The search strategy consisted of four components. First, we manually selected the term "raynaud" to retrieve all relevant articles containing a source term in the title and/or abstract fields (`TIAB`). Second, we applied a date range constraint to match the same period as in Swanson's work (`PDAT`). Third, only high-quality articles published in the MEDLINE subset of PubMed were retrieved (`SB`). The fourth component included articles written only in English (`LA`).

[1] https://arrowsmith.psych.uic.edu/arrowsmith_uic/data/stopwords_swanson

Table 7.1 Raynaud's disease–fish oil pathways with corresponding b-concepts.

Blood viscosity	Platelet aggregation	Vascular reactivity
Blood viscosity	Fibrinolysis	Vasodilatation
Erythrocyte deformability	Platelet aggregation	Vasospasm
Plasma viscosity level	Thrombosis	Vasospasm mechanisms
Hemorheology	Platelet adhesiveness	Vasomotion
Decreased vascular flow	Effects, blood coagulation	Decreased vascular resistance
Hyperviscosity		Decreased vascular flow

To normalize free text from the collected titles and abstracts, we used the MetaMap tool from the NLM (Aronson & Lang, 2010). MetaMap extracts biomedical concepts (e.g., Raynaud Disease) from raw text and then maps them to concept unique identifiers (CUIs) in the UMLS Metathesaurus. For example, for the input title "Fish oil, Raynaud's syndrome, and undiscovered public knowledge", MetaMap first extracts the following three phrases:

1. Fish oil
2. Raynaud's syndrome
3. undiscovered public knowledge

The next stage retrieves the candidate concepts for each noun phrase from the UMLS based on a word match between the concept name and the text. The MetaMap begins with the parsing of a document into sentences, followed by the segmentation of each sentence into phrases and further division of each phrase into individual words. Candidate concepts for each noun phrase are obtained from the ontology by comparing the terms in the concept name with the text. For example, the phrase "Raynaud's syndrome" is mapped to the UMLS concept "raynaud syndrome (Raynaud Phenomenon)" associated with semantic type Disease or Syndrome. The phrase "undiscovered public knowledge" is mapped to the UMLS concept "Knowledge" with semantic type Intellectual Product. Raynaud's disease–fish oil pathways with corresponding metamapped b-concepts are listed in Table 7.1.

7.2.2 Knowledge-based approaches to reduce the concept space

The usual strategy in knowledge-based LBD approaches is to reduce the concept space to only those concepts linked to a predefined set of semantic types. For example, if interested only in non-drug related approaches for treating the diseases in question, we can only select UMLS concepts that represent non-drug approaches.

To reduce the search space during the LBD process, we usually filter UMLS-mapped concepts using semantic types through a technique called *semantic filtering*. To better understand this semantic filtering approach, inspect a selection of semantic resources

Table 7.2 Fish oil *a* concepts with ranks for each pathway in the open discovery task.

Blood viscosity	Platelet aggregation	Vascular reactivity
20 Fish oils	50 Eicosapentaenoic acid	56 Fatty acids, essential
21 Maxepa	54 Cod liver oil	57 Dietary fats
23 Fatty acids, omega-3	77 Fish oils	
26 Omega-3 fatty acid	94 Maxepa	
32 Eicosapentaenoic acid	135 Fatty acids, omega-3	
54 Epa-e	155 Omega-3 polyunsaturated fatty acid	
	163 Salmon oil	

used in LBD research described in Section 2.4.2, including the UMLS Semantic Network, which is one of the three main knowledge sources incorporated in UMLS.[2]

For example, in the stage of selecting bridging concepts (*b*-terms), we might select only those concepts that have a functional semantic type, such as "Biologic Function", "Cell Function", and "Phenomenon or Process". Additionally, open and closed discovery modes may employ different semantic filters. A restrictive filter is necessary for an open discovery mode, which will yield a large number of PubMed records (and extracted knowledge concepts), while a less restrictive filter may be more beneficial for a closed discovery mode. Table 7.2 summarizes the corresponding *a*-concepts together with their ranks in a reduced concept space for the fish-oil domain. Semantic filtering is usually performed using a human-in-the-loop approach during the discovery process and requires domain expertise to select the optimal set of semantic types.

An alternative to semantic filtering is the *semantic profiling* approach. The technique of semantic profiling is based on the notion of topic profile (TP), which in turn is a semantic type vector. Each semantic type vector consists of weighted MeSH headings that belong to a specific semantic type. Since there are 134 semantic types in the UMLS ontology, each TP is formed from 134 semantic type vectors. For example, in Equation (7.1), $w_{i,134,1}$ refers to the weight of the MeSH descriptor $m_{134,1}$, which in turn is associated with the 134th semantic type.

$$\mathrm{TP}_i = [\{w_{i,1,1}m_{1,1}, w_{i,1,2}m_{1,2}, \dots\}, \dots, \{w_{i,134,1}m_{134,1}, \dots\}] \tag{7.1}$$

The MeSH weight is determined by a TF-IDF technique, with a TP indicating the relative significance of the different semantic types. van Haagen et al., 2009 later introduced the *concept profile*, a method similar to the TP technique, which was designed to measure the similarity between protein concepts.

[2] https://lhncbc.nlm.nih.gov/semanticnetwork/

7.3 RaJoLink open discovery methodology

In this section, the RaJoLink method (Petrič et al., 2009) is presented as a specific implementation of the open LBD approach. The procedural steps and underlying principles of RaJoLink are explained in detail and illustrated with some practical cases that demonstrate how to navigate the vast landscape of published research to discover new links within the available scientific literature.

Compared to other LBD approaches, the main novelty of the RaJoLink methodology is its focus on identifying *rare terms* in a domain literature C with a potential for new discoveries. By focusing on rare terms, RaJoLink differs from conventional models by proposing a novel method for generating association candidates, thereby expanding the range of possible bisociative discoveries beyond the standard closed discovery LBD approaches. RaJoLink employs a structured process that starts with the identification of these rare terms that are considered outliers within the domain-specific literature corpus C. These rare terms act as potential intermediate concepts b in the search for a related domain A in open discovery. To distinguish these rare terms used in open discovery from the bridging terms b_i considered in the closed discovery process, we will denote the rare terms with r_i. Actually, several rare terms r_i are considered, each defining a potential domain of interest for open discovery. The method then examines the literatures R_i on these rare terms r_i, and looks for commonalities through shared terms a_i in the intersection of these domains $\bigcap_i^m R_i$, among which the most suitable joint term a defining the candidate domain A would be selected by the expert. Finally, domain A and the originally specified domain C are used in closed discovery to find interesting bridging terms b_i.

The idea of combining open and closed discovery is illustrated in Figure 7.2, where terms r_i are the rare intermediate concepts identified for open discovery, while b_i stand for the bridging terms identified in closed discovery.

This approach thus builds a connection between seemingly unrelated areas, which is facilitated by the candidate domains R_i resulting from rare terms r_i. The main rationale for selecting candidates for a through literatures R_i on rare terms r_i are as follows. If there are some rare terms occurring in the literature on C, one should first check the available records (published scientific papers) on these rare terms. If these records contain an interesting common/joint term a in the intersection of these papers (i.e. in $\bigcap_i^m R_i$), one should further check whether it is logically related to C. An inherent assumption in this process is that focusing on rare terms might increase the likelihood that the proposed candidates have not yet been analyzed for their connections to C.

Although rare terms (or outliers) in open LBD may lead to innovative findings and new connections, they might also pose a major challenge. Researchers must be cautious and apply rigorous methods and critical thinking to distinguish between outliers that represent meaningful deviations worthy of further investigation and those that represent noise or error. Successfully overcoming these challenges is critical to realize the full potential of outliers to gain scientific insight (Aggarwal, 2013).

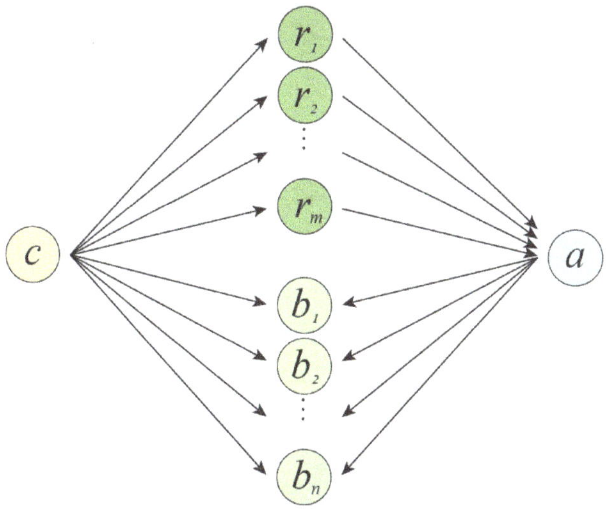

Fig. 7.2 Combining open and close discovery in RaJoLink. In open discovery, only the starting concept c is given and the goal is to identify rare terms r_i, whose analysis should lead to a the identification of the most interesting rare term as target concept a to define target domain A. Domain A is then used for closed discovery to find interesting b-terms $b_1, b_2, \ldots, b_n$, representing bridges between A and C.

7.3.1 Semantic filtering in RaJoLink

Standard data preprocessing in RaJoLink includes word lemmatization and the exclusion of words from a stoplist. A stoplist consists of stop-words, i.e. words that are predictably uninteresting and can therefore be excluded from the input datasets. In this way, all terms that are not topic-related can be ignored. We use a list of 571 English stop-words (i.e. more general words such as a, able, about, above and according).

In addition, in the Ra and Jo steps of the RaJoLink methodology, text preprocessing involves also semantic filtering in order to make the search for rare terms effective. Filtering according to the MeSH classification (Nelson et al., 2001) is used. The method presented here first compares up to three-word terms from the input text with the MeSH terms. If a multi-word term is found among the MeSH terms, it is added to the list of terms for further statistical analysis of the strings. The same is then done with two-word terms from the input text. Multiword terms that are not found in the MeSH thesaurus are treated as single words. There are several top-level nodes in the MeSH tree: Anatomy [A], Organisms [B], Diseases [C], Chemicals and Drugs [D], Analytical, Diagnostic and Therapeutic Techniques, and Equipment [E], Psychiatry and Psychology [F], Phenomena and Processes [G], Disciplines and Occupations [H], Anthropology, Education, Sociology, and Social Phenomena [I], Technology, Industry, and Agriculture [J], Humanities [K], Information Science [L], Named Groups [M], Health Care [N],

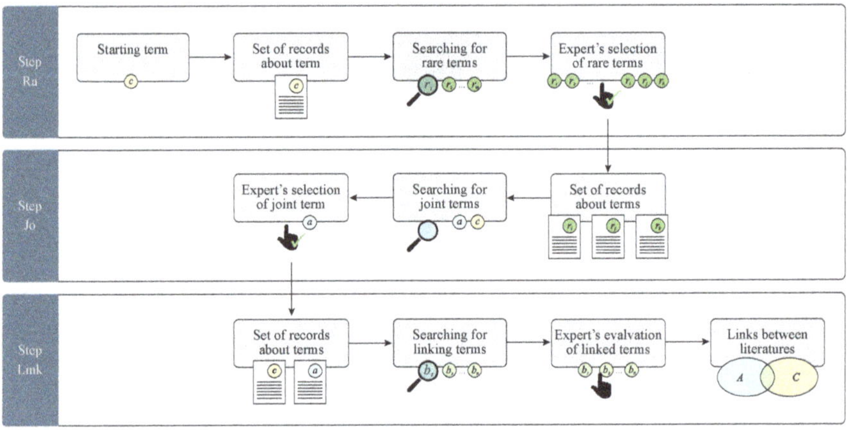

Fig. 7.3 RaJoLink methodology workflow showing the procedures of steps Ra, Jo and Link.

Publication Characteristics [V] and Geographicals [Z]. Since in our exploration the first domain was typically selected from Diseases [C], we limited the search for second linking rare words to the Chemicals and Drugs [D]branch, more specifically Enzymes and Co-enzymes [D08] and Amino Acids, Peptides, and Proteins [D12]. It can be assumed that words that are of interest to medical science are included in the Medical Subject Headings. Concentrating on such words can narrow down the search space and thus speed up and improve the discovery process.

7.3.2 RaJoLink methodology

The RaJoLink methodology, proposed by Petrič et al. (2009), consists of three main steps, Ra, Jo, and Link, named after the key elements of each step: *rare terms*, *joint terms*, and *linking terms*. The open LBD approach is implemented in the Ra and Jo steps of the RaJoLink method, while the Link step implements close discovery. Note that in the Ra and Jo steps of the RaJoLink methodology, document representation using MeSH terms is used, while the standard Bag of Words (BoW) document representation is used in the Link step of the methodology.

The RaJoLink methodology, combining the open and closed discovery approach, can be summarized as follows. In step Ra, interesting rare terms are identified in the literature about the studied phenomenon C. In step Jo, all available articles on the selected rare terms are examined and interesting joint terms that appear in the intersection of the literature on the rare terms are identified as candidates for A. This results in a candidate hypothesis that C is related to A. To provide an explanation for the hypotheses generated in step Jo, in the Link step the method searches for b-terms that connect the literature on joint term a (literature A) with the literature on term c (literature C). The overall methodology is illustrated in Figure 7.3.

Note that in the Ra step of the RaJoLink methodology, the implemented outlier-based approach is founded on the principle of rarity. The principle of rarity has been previously studied in the field of ecology statistics (Ellison & Agrawal, 2005). These studies included the rarity driven by biodiversity and conservation policy (Carney, 1997). In light of this, ecologists have been particularly concerned with the study of rare species (Boughton, 2001). They recognized two syndromes of rarity: habitat-limited species that are rare because their habitat is rare, and distribution-limited species that are rare because they have been left behind by a catastrophic upheaval of old growth. While the main concern of ecologists was to prevent the extinction of rare species, they also recognized the potential of dispersal-limited species to adapt to a changing environment.

The individual steps of the RaJoLink methodology are outlined below.

Step Ra. The goal of the Ra step is to identify rare information in a set of documents on the term c (literature C) to increase the chance of discovering useful implicit relations that have not yet been published in the literature.

Let us denote the total number of records in the input file by N and by $n(T)$ the number of documents containing the term T. As mentioned in Section 7.3.1, in the Ra step of the methodology, the terms considered are limited to the MeSH terms. In practice, we focus on terms that rarely appear in the input data set. A term appears rarely in the input set of records if it occurs in a relatively small part of the input set. Let us assume that the term T is rare if $n(T)$ is equal to 1. Note, however, that such a restriction is very sensitive to the addition of new elements to the input set of records, as the rarity/frequency of a term in a text corpus can change by adding new texts to the existing input corpus. A rare term may become more frequent if the text added to the input file contains this term.

For each MeSH term T that appears in the input datasets, $n(T)$ is calculated using frequency statistics. As already mentioned, all terms with $n(T)$ equal to 1 are selected as rare. Let us denote such terms by r. A domain expert then selects a few interesting rare terms $r_1, r_2, ...r_p$ considering the background knowledge about the topic of interest. Note that p can be considered as a parameter of the method. If rare terms cannot be found with this minimum frequency, the experimenter must return to the beginning of the Ra step and repeat the process by considering a higher threshold $n(T)$ for the number of input documents. However, the appearance of selected rare terms in the literature C means that they have already been reported in connection with the term c. The rare terms therefore only represent intermediate results on the way to the discovery of new knowledge. It should also not be assumed that they have the same properties as b-terms b, although the rare terms are individually associated with literature A and literature C, but not with A and C together. The reason for this is that when testing the hypotheses by searching for meaningful linking terms b between literatures A and C, the focus should be on the most frequent terms, while in the hypothesis generation phase the rarest terms are of interest.

Step Jo. When step Ra is completed, new individual data sets, one for each selected rare term, must be obtained and further analyzed. For this purpose, a set of articles for each of the rare terms selected in step Ra is automatically retrieved from MEDLINE or extracted from another document source. After preprocessing each set of articles as in step Ra, the goal is to find the terms that the text collections have in common. A term a_k is considered a joint term if it occurs in at least two datasets R_i defined by the rare terms r_i; this constraint ensures that the identified text collections A_k defined by the identified terms a_k are not completely disconnected from each other. At the same time, the requirement is that the joint term a_k must not occur in the set C of records defined by the term c created in step Ra; this constraint ensures that a_k has not yet been considered in the literature C. Thus, the intermediate result of this step are joint terms $a_1, a_2, \ldots, a_q$ at the intersection of the literatures R_i on rare terms. Among these terms a_i, the expert selects the term that is considered the most promising for further exploration. This term a_i, denoted as a, defines the domain A used in the subsequent closed discovery step Link.

If through the above-described literature exploration in step Ra one does not obtain joint terms, the user needs to return to the results of the previous step and expand or change the selection of rare terms and repeat the process. The role of the expert is crucial in step Jo. Based on the expert's opinion, one of the suggested joint terms a_k is selected for further investigation in step Link. The selected joint term is renamed as a to define domain A, and used in the search for new hypotheses according to Swanson's ABC model for closed discovery in step Link.

Step Link. As opposed to data representation in steps Ra and Jo (which focused on exploring MeSH terms), in the Link step the documents are represented using the Bag of Words (BoW) representation, using the standard TF-IDF (Term Frequency - Inverse Document Frequency) word weighting method (Salton & Buckley, 1988). Text processing and document similarity computation are performed within the OntoGen tool (Fortuna et al., 2006), where cosine similarity (see Equation (3.2)) is used as a measure of document similarity.

To provide an explanation for the hypotheses generated in step Jo, the RaJoLink method looks for connections between the literature on the joint term a and the literature C defined over the original term c. This step of the methodology is implemented using the closed discovery approach described in Section 6.3. Again, the approach used in the Link step includes the unique aspect of our method compared to the approaches studied by other LBD researchers, as the discovery of b-terms is focused on outliers $O(A)$. In this approach, already mentioned in Section 6.3 and illustrated in Figure 6.5, the most interesting document considered for exploration are the documents from A that are ones that are the closest to the centroid document $c \in C$ in the similarity graph, defined over the documents of combined dataset of literatures $C \cup A$. In practice, within the entire corpus of the text dataset that consists of literatures A and C, each text document represents a single record. All records are then sorted according to their similarity to the centroid c of document set C, and the similar documents are inspected by comparing adjacent documents from the list. Since searching for the linking terms can be very complex, focusing on neighboring documents can be used as a heuristic guide to speed up the discovery process by the expert.

7.3.3 Application: Finding connections between autism and calcineurin

We reviewed the literature on autism, a neurodevelopmental disorder that affects approximately 1 in 100 children (Zeidan et al., 2022). The Diagnostic and Statistical Manual of Mental Disorders classifies autism as a pervasive developmental disorder with symptoms of neurological development, associated with early brain mechanisms (American Psychiatric Association, 2000). Characteristic features include impaired social interactions, communication problems and restricted, repetitive patterns of behavior (Georgiades et al., 2007). The etiology of autism remains elusive, with current research suggesting a mixture of genetic, environmental and neurological factors.

Recent advances have shed light on the neurological discrepancies of autism, such as atypical neuroanatomy and cellular neurochemistry (Bauman & Kemper, 2005; DeVito et al., 2007). The results suggest that various molecular alterations in brain development and neurotransmission, abnormal neuron morphology and changes in brain volume contribute to the complexity of autism. This neurological diversity underscores the potential variability in the causes of autism. Our study focused on exploring the biochemical and neurological basis of autism and provided important insights into the neural mechanisms involved, which play a central role in understanding the pathology of autism.

In our study, we exploited the PubMed search engine to examine the autism literature using the RaJoLink method. PubMed, a gateway to the MEDLINE database and other major bibliographic resources, allowed us to search for representative 214 full text documents from the decade prior to our experiment in the autism domain. After gathering this comprehensive document collection, we carefully performed preprocessing, converting the documents from HTML and PDF formats to plain text and ensuring that each document was consistently formatted for further analysis. Despite the challenges of accessing and processing full-text articles, we were able to extract from these texts around 2000 unique terms, with a particular focus on rare terms from amino acid, peptide, and protein domains which we tested for their potential relevance to autism research. Notable rare terms such as *lactoylglutathione*, *synaptophysin* and *calcium channels* appeared in our dataset, leading the autism expert of our team to specifically investigate their similarities and associations with calcineurin (since it appeared as a joint term in all literatures defined by the selected rare terms)—a protein phosphatase with high prevalence in the brain (see Figure 7.4).

The link step of the RaJoLink method facilitated the discovery of significant connections between autism and calcineurin, bringing to light terms such as Bcl-2, calmodulin and synaptic plasticity. These terms provide a bridge between autism and broader biological processes and underscore the utility of the method in identifying new avenues of research. The results highlight the unique overlaps between the autism and calcineurin literature and point to potential areas for further investigation.

Another concept was also suggested for exploration: Fragile X syndrome. Extending our research to datasets that mention autism and Fragile X syndrome using the RaJoLink method brought to light new relevant links, such as the identification of NF-kappaB as a potential player in autism. This underlines the method's ability to uncover unexpected connections that could be relevant for future research directions.

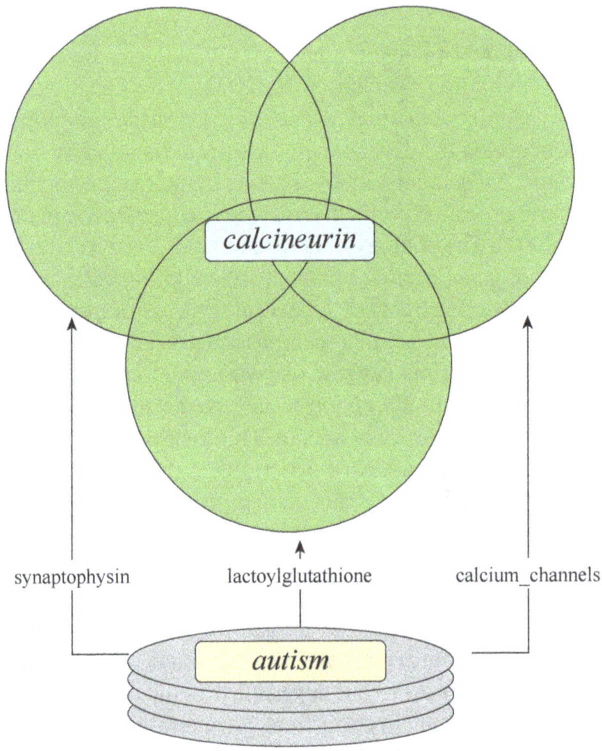

Fig. 7.4 From autism to calcineurin: results obtained with RaJoLink open discovery on autism domain.

The biomedical significance of using RaJoLink Open Discovery goes far beyond autism research. This method represents a major shift in the way medical science can uncover hidden connections in large amounts of literature and improve the understanding of complex diseases. In this way, by linking disjoint datasets, RaJoLink enabled a deeper understanding of some of the underlying biological mechanisms, which can lead to novel therapeutic targets and diagnostic markers.

The qualitative assessment of the hypothesis linking calcineurin and autism was performed by a specialist physician who confirmed the suggested connections between two distinct under-researched areas. This validation was based on recent findings, highlighting the role of calcineurin in intracellular signaling pathways that control synaptic plasticity and neuronal activity. In addition, the expert drew parallels with Fragile X syndrome and pointed out that impaired synaptic plasticity could be due to the absence of the FMR1 protein, which is known to contribute to autism. Despite the absence of direct evidence for a link between calcineurin and autism phenomena found online, the proposal to investigate the role of calcineurin in autism, which at the time of this experiment represented a compelling avenue for research.

Human involvement proved to be crucial in selecting rare and common terms to tailor the search to relevant and significant results in the Ra and Jo steps of the methodology, while the Link closed discovery step reduced the complex combinatorial search for linking terms by prioritizing terms from closely related documents, enabling a more efficient expert selection process. The RaJoLink approach thus accelerated the discovery of valuable terms for further investigation.

Expanding the use of RaJoLink open discovery approach in biomedical research, particularly in the field of autism, may lead to further insights and a deeper understanding of the multifaceted nature of autism. With the increasing complexity of biomedical data, tools like RaJoLink may become essential in the search for knowledge and the development of innovative medical treatments.

7.3.4 Discussion

The growing body of biomedical literature represents both an opportunity and a challenge for the discovery of novel hypotheses to solve unsolved problems in medical science. This vast landscape of texts can often lead to information overload, making the extraction of valuable knowledge a daunting task. To help researchers navigate this complexity, we have introduced the RaJoLink method, which relies on word frequency statistics and semantic text analysis techniques based on Swanson's ABC model. Unlike traditional methods that start from a predefined hypothesis, RaJoLink emphasizes open discovery and leads researchers to new hypotheses without preconceived notions.

The novelty of the RaJoLink method lies in its emphasis on the principle of rarity in the open literature-based discovery setting. By focusing on rare terms in the biomedical literature, especially those related to topic c, we can direct the search to unexplored hypotheses. This approach is further enhanced by the use of bisociations, a concept that goes beyond simple associative information retrieval by enabling the discovery of relationships across different contexts. Such bisociative connections are crucial for fostering innovative discoveries that are often overlooked in traditional associative methods.

Our implementation of the RaJoLink method in the context of MEDLINE articles has emphasized its potential to uncover bisociative relationships that are critical to creative breakthroughs in medical research. By examining rare terms that serve as overlaps between different contexts, the method facilitates the identification of novel connections, as demonstrated in our study of the links between calcineurin and autism. This methodological innovation enables the comprehensive analysis of entire article texts and thus differs from the approaches of Swanson or Weeber, who relied mainly on the titles and summaries of articles.

The discovery of significant links between calcineurin and autism literatures by RaJoLink demonstrates the effectiveness of the method in bridging disparate fields to generate new knowledge. Such discoveries are critical to autism research and biomedical science in general, as they illuminate previously unrecognized pathways and mechanisms that can contribute to the understanding and treatment of complex disorders.

Collaboration with experts improves this process and enables the rapid translation of discovered connections into meaningful biomedical insights.

Exploitation of outliers in RaJoLink is beneficial in two stages: (1) the use of rare terms (outliers) to search for links in the open discovery (steps Ra and Jo in RaJoLink) from C to A, and (2) reduction of the input data volume by concentrating on outlier documents (OntoGen and classified as C) in the closed discovery. In addition, the innovative use of rare terms to develop new hypotheses represents a significant advance in literature-based biomedical research. However, in specific research scenarios the effectiveness of the RaJoLink open LBD approach may be limited due to its heavy reliance on the identification of rare terms in the domain-specific literature. Although focusing on rare terms can uncover new insights and connections, there is a risk that valuable information contained in frequently discussed terms or topics can be neglected and important associations that are well documented but not rare could be overlooked, resulting in incomplete or biased research results.

Another issue is the possibility of a high false positive rate in the hypotheses generated by RaJoLink, since trying to generate numerous links from sparse data increases the likelihood of identifying irrelevant or erroneous links. This could steer researchers towards less productive paths or require additional review and validation processes to sift through the hypotheses produced. The open-ended nature of the method, while beneficial for exploratory research purposes, requires rigorous subsequent testing and validation procedures that can be resource- and time-consuming, potentially negating the efficiency gains in the initial discovery phase. Finally, note that the implementation and realization of the RaJoLink approach can be quite challenging, especially when it comes to sufficient computing resources and knowledge. The approach relies on access to comprehensive and up-to-date databases such as MEDLINE. For researchers with limited resources or lacking the necessary technical skills to take full advantage of these tools, the benefits of RaJoLink are limited. In areas where research materials are not fully digitized or are not easily accessible, RaJoLink's ability to develop trustworthy and innovative hypotheses may be weakened.

7.4 Tutorials in Python

The content presented in this chapter is supplemented by a practical example, for which the source code is hosted on GitHub. Readers are encouraged to access the tutorials **02_open_discovery** and **05_mini_rajolink** at the book's GitHub repository to further explore the concepts discussed. Before executing the provided Jupyter notebooks, the readers are advised to read Chapter 4 and follow the instructions to install Python and the required packages from **requirements.txt** file included in the repository.

7.4.1 02_open_discovery

In this tutorial, we extend the closed discovery approach to the open discovery mode to generate new research hypotheses. We guide the reader through reproducing the second Swanson's (1988) landmark discovery, which connects migraine headaches to magnesium, aiming to identify novel therapeutic candidates (A) for migraine headaches (C).

Rather than relying on phrases extracted from plain text, this tutorial leverages a structured knowledge base, an approach originally introduced by Weeber et al. (2001). However, in the present implementation, we use the Medical Subject Headings (MeSH) ontology to facilitate the LBD process. The use of MeSH allows for structured background knowledge to reduce textual variations and synonyms of terms to standardized knowledge concepts.

Each document in the dataset is represented as a list of MeSH headings. The LBD process begins with a dataset of 1156 PubMed articles on migraines, which map to a total of 1812 unique MeSH headings. To narrow the search space, we apply a semantic type filter, restricting MeSH headings to categories such as "Amino Acid, Peptide, or Protein", "Pathologic Function" and "Phenomenon or Process", resulting to a more manageable set of 116 MeSH headings.

Next, we rank the filtered MeSH headings using TF-IDF and select three b-concepts for further investigation: "Vasoconstriction", "Platelet Aggregation", and "Spreading Cortical Depression". For each b-concept, we perform a new PubMed query to identify associated a-concepts. This search retrieves 2898, 6273, and 180 bibliographic records for "Vasoconstriction", "Platelet Aggregation", and "Spreading Cortical Depression", respectively. Applying the semantic type filter to these records yields 11, 73, and 13 unique MeSH headings for the respective b-concepts.

The intersection of these results contains ten candidate a-concepts. After applying rank aggregation, two concepts, "Calcium" and "Magnesium", emerge as the top-ranked candidates. Notably, "Calcium" already has four direct connections to migraines in the literature, whereas "Magnesium" shows no prior joint reports, making it a promising novel hypothesis.

7.4.2 05_mini_rajolink.ipynb

RaJoLink represents an outlier-based open-discovery three-step method, applied to the autism dataset. The method requires a pre-defined list of terms. In the first step (Ra), we limit the terms to MeSH headings, corresponding to semantic types "Enzymes and Coenzymes" and "Amino Acids, Peptides, and Proteins" (Lipscomb, 2000), resulting in 495 terms of interest. After that, TF-IDF weighting of the terms is conducted across documents, resulting in TF-IDF BoW representation over the retrieved MeSH terms. Next, the representation is aggregated per term, resulting in a single score per term. Finally, an expert examines the terms based on their positioning over the aggregated BoW representation, selecting the most interesting ones. This manual intervention identified three terms out of 495, specifically calcium channel, synaptophysin, and lactoylglutathione, for further investigation of their association with autism. The three selected rare terms are in positions 38, 37 and 377 of in the sorted list of candidate rare terms.

In the next step (Jo), the documents (titles and abstracts) for each rare term were retrieved from PubMed and combined into a single document list. 11,666 documents from all three domains were collected and then pre-processed similarly to step Ra. The BoW model was created and aggregated by domain. Based on the BoW model, 428 common word candidates were selected and sorted according to the TF-IDF scores. The expert selected the term *calcineurin*, which was in position 18 of the list of 428 sorted joint word candidates.

The last step Link in RaJoLink implements the principle of closed discovery between two domains, in our case *autism* and *calcineurin*. The notebook `05_mini_rajolink.ipynb` illustrates the three-step RaJoLink LBD procedure used in the open LBD setting by Petrič et al., 2012, using autism as the start domain of interest.

References

Aggarwal, C. (2013). *Outlier Analysis*. Springer.

American Psychiatric Association. (2000). *Diagnostic and statistical manual of mental disorders, fourth edition.*

Aronson, A. R., & Lang, F.-M. (2010). An overview of metamap: Historical perspective and recent advances. *Journal of the American Medical Informatics Association, 17*(3), 229–236. https://doi.org/10.1136/jamia.2009.002733

Bauman, M. L., & Kemper, T. L. (2005). Neuroanatomic observations of the brain in autism: A review and future directions. *International Journal of Developmental Neuroscience*, 183–7.

Boughton, D. (2001). Paradoxes in science: A new view of rarity. *Sci. Findings, 35*, 1–6.

Carney, R. S. (1997). Basing conservation policies for the deep-sea floor on current-diversity concepts: A consideration of rarity. *Biodiversity and Conservation, 6*(11), 1463–1485.

DeVito, T. J., Drost, D. J., Neufeld, R. W., Rajakumar, N., Pavlosky, W., Williamson, P., & Nicolson, R. (2007). Evidence for cortical dysfunction in autism: A proton magnetic resonance spectroscopic imaging study. *Bilological psychiatry*, 465–73.

Ellison, A. M., & Agrawal, A. A. (2005). The statistics of rarity. *Ecology, 86*, 1079–1080.

Fortuna, B., Grobelnik, M., & Mladenić, D. (2006). Semi-automatic data-driven ontology construction system. *Proceedings of the 9th International Multi-conference Information Society*, 223–226.

Georgiades, S., Szatmari, P., Zwaigenbaum, L., Duku, E. K., Bryson, S. E., Roberts, W., Goldberg, J., & Mahoney, W. J. (2007). Structure of the autism symptom phenotype: A proposed multidimensional model. *Journal of the American Academy of Child and Adolescent Psychiatry*, 188–96.

Lipscomb, C. E. (2000). Medical Subject Headings (MeSH). *Bulletin of the Medical Library Association*, *88*(3), 265–266.

Nelson, S. J., Johnston, W. D., & Humphreys, B. L. (2001). Relationships in Medical Subject Headings (MeSH). In C. A. Bean & R. Green (Eds.), *Relationships in the organization of knowledge* (pp. 171–184). Springer.

Petrič, I., Cestnik, B., Lavrač, N., & Urbančič, T. (2012). Outlier detection in cross-context link discovery for creative literature mining. *The Computer Journal*, *55*(1), 47–61.

Petrič, I., Urbančič, T., Cestnik, B., & Macedoni-Lukšič, M. (2009). Literature mining method RaJoLink for uncovering relations between biomedical concepts. *Journal of Biomedical Informatics*, *42*(2), 219–227.

Salton, G., & Buckley, C. (1988). Term-Weighting Approaches in Automatic Text Retrieval. *Information Processing & Management*, *24*(5), 513–523.

Swanson, D. R. (1986a). Fish oil, Raynaud's syndrome, and undiscovered public knowledge. *Perspectives in Biology and Medicine*, *30*(1), 7–18.

Swanson, D. R. (1986b). Undiscovered public knowledge. *The Library Quarterly*, *56*(2), 103–118.

Swanson, D. R. (1988). Migraine and magnesium: Eleven neglected connections. *Perspectives in Biology and Medicine*, *78*(1), 526–557.

van Haagen, H. H., 't Hoen, P. A., Bovo, A. B., de Morrée, A., van Mulligen, E. M., Chichester, C., Kors, J. A., den Dunnen, J. T., van Ommen, G. J. B., van der Maarel, S. M., Kern, V. M., Mons, B., & Schuemie, M. J. (2009). Novel protein-protein interactions inferred from literature context. *PLoS ONE*, *4*(11).

Weeber, M., Klein, H., de Jong-van den Berg, L., Vos, R., et al. (2001). Using concepts in literature-based discovery: Simulating Swanson's Raynaud–fish oil and migraine–magnesium discoveries. *Journal of the American Society for Information Science and Technology*, *52*(7), 548–557.

Zeidan, J., Fombonne, E., Scorah, J., Ibrahim, A., Durkin, M., Saxena, S., Yusuf, A., Shih, A., & Elsabbagh, M. (2022). Global prevalence of autism: A systematic review update. *Autism Research*, *15*(5), 778–790.

Chapter 8
Network-based Closed Discovery

Link prediction in complex networks aims to infer potential future connections between the nodes by analyzing the network's current structure and its past evolution. Over the past decade, this challenge has attracted significant research interest across different scientific fields, including LBD.

Complex networks are dynamic evolving structures that change over time by adding new nodes or by forming new connections between the existing nodes. In the context of knowledge representation, these nodes can be understood as individual concepts or pieces of information, whereas the links between them represent the established relationships, associations, or interactions that connect these concepts.

In this chapter, we study link prediction in the context of closed LBD. Section 8.1 presents the background for using network analysis and link prediction for LBD. Section 8.2 presents a specific link prediction approach to LBD. The chapter concludes with a discussion in Section 8.3, followed by presenting selected sources and code reuse in Section 8.4.

8.1 Co-occurrence and link prediction background

The prediction of various forthcoming phenomena is a general scientific challenge and is of utmost importance in practically all disciplines. Recent years have seen an increasing interest in the study of large, real-world complex networks, in which graph theory is used to model the relationships among the data entities. Under the umbrella of text mining, researchers are close to processing the entire available literature to extract biological interactions (Li et al., 2013). Knowledge of a particular biomedical domain can be viewed as an instance of such a complex network, which is composed of a set of biomedical concepts and the relationships among them (Bales & Johnson, 2006). For example, in pharmacogenomics, relationships among genes, diseases, and chemical substances constitute an important part of biomedical knowledge.

Associations between entities based on the co-occurrence of biomedical terms, such as chemical substances, biological processes, and diseases or genes constitute a knowl-

N. Lavrač et al., *Bisociative Literature-Based Discovery*,
https://doi.org/10.1007/978-3-031-96863-1_8

edge representation. A co-occurrence approach is built on the assumption that biomedical concepts co-occurring in the same title or abstract are in some way biologically related (Cohen & Hunter, 2008). Simple co-occurrence of concepts can be extended by the number of times a concept is found in a document or by the closeness between one concept and another concept in a sentence (Alako et al., 2005). Many applications have been developed, exploiting co-occurrence to mine interesting patterns in biomedical resources (Hoffmann & Valencia, 2005; Thomas et al., 2012).

These associations can be represented as a graph consisting of nodes and edges, where the former represent concepts and the latter relationships. Contemporary network analysis is a suitable framework for the analysis of such data structures.

Link prediction is a research direction at the intersection of network analysis and machine learning. Link prediction refers to the discovery of associations between nodes that are not directly connected in the current snapshot of a given network, but will be connected in the future (Martinez et al., 2016). Link prediction in complex networks aims to estimate the probability that an edge exists between two nodes, based on the observation of existing edges and the attributes of the nodes (Yu et al., 2014).

Several techniques have been proposed to predict new edges on the basis of the observed network topology. We refer the reader to two papers for an exhaustive introduction to link prediction (Kumar et al., 2020; Martinez et al., 2016).

Link prediction algorithms can be classified into two groups: unsupervised and supervised (Martinez et al., 2016). In unsupervised learning, the probability of connection is measured by a similarity metric between the two nodes. Local similarity metrics depend only on the information of the neighborhoods, such as the common neighbors or the Jaccard coefficient. Global similarity metrics require the entire topological information of the network from the perspective of paths or random walks. Unsupervised algorithms do not include learning, but rather they employ predefined scores that are invariant to the specific structure of the input network. On the other hand, in supervised learning some pattern of the link behavior based on multiple unsupervised scores represented as a feature vector is introduced and used as input to supervised learning algorithms, such as decision trees or support vector machines.

8.2 Link prediction approach to LBD

This section studies link prediction from the perspective of LBD, and presents a specific link prediction approach to LBD, which complements more traditional text-based LBD approaches discussed in Chapter 5. The section builds on our earlier work (Kastrin et al., 2014c) where we presented the basic idea of LBD using the link prediction methodology, whose specific goals are two-fold:

1. To formulate LBD as a classification problem, where features are represented by different topological similarity indices between biomedical concepts.
2. To evaluate the performance of unsupervised and supervised link prediction methods for predicting implicit or previously unknown links between biomedical concepts.

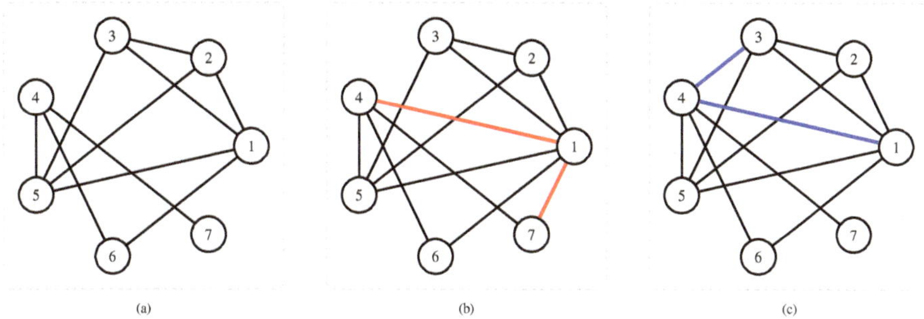

Fig. 8.1 Link prediction in a complex network. New link formation is predicted (b) from the topology of a training network (a) and evaluated on an independent test network (c).

This section starts by presenting the experimental setup, followed by the presentation of the link prediction methodology, performance evaluation and results. To evaluate link prediction techniques for LBD, we investigate link prediction performance on a complex network obtained from the MeSH terminology.

8.2.1 Experimental setup

We processed the entire MEDLINE Baseline Repository, up to the end of 2010, which contained 21 850 751 records tagged with MeSH descriptors (MeSH terms). As the distribution is in XML format, we extracted the relevant elements and transformed them into a relational text format (i.e. one line for each major MeSH term occurrence in each record). To set up a co-occurrence network, we considered each MeSH term as a node. An edge between two MeSH terms was defined if they appear together in the same MEDLINE record. Note that we did not consider the direction of the relations (i.e. the relation between MeSH terms u and v is the same as the relation between terms v and u), that is, the edges are undirected. The network was represented in the edge list form.

As mentioned in Section 3.4.1, a network is represented by a graph $G(V,E)$ that consists of a set of nodes V representing major MeSH terms, and a set of edges E representing relationships between the nodes. Suppose that multiple edges and loops are not allowed. Let us denote by U the universal set of all possible edges in G. The set of non-existing edges is then denoted by $U \setminus E$. Next, we propose that there are some missing edges in the set $U \setminus E$. The task of link prediction is to discover these missing edges (Lü & Zhou, 2011). The link prediction task is illustrated in Figure 8.1.

To solve the link prediction problem, we need to understand the formation dynamics of edges between all possible pairs of nodes. The link prediction problem can be formally represented as follows (Liben-Nowell & Kleinberg, 2003). Let us suppose having a network $G[t_1,t_2]$, which contains all interactions among nodes that take place in the time interval $[t_1,t_2]$. Next, suppose that $[t_3,t_4]$ is a time interval occurring after $[t_1,t_2]$. The task of link prediction is to provide a list of edges that are present in

$G[t_3, t_4]$ but are absent in $G[t_1, t_2]$. We refer to $G[t_1, t_2]$ as the training network and $G[t_3, t_4]$ as the testing network. The former is composed of the set of edges E^T and the latter of the set of edges E^P, such that $E^T \cup E^P = E$ and $E^T \cap E^P = \emptyset$.

For each pair of nodes in the training network we could compute various statistics (e.g., similarity measures), which reflect the probability that a pair of nodes will co-occur in the testing network. Each such pair represents a positive or negative instance, depending on whether these node pairs form at least one edge in the test interval.

8.2.2 Link prediction methodology

Link prediction was performed using various proximity measures that are used to find the similarity between a pair of nodes. Our approach relies on the assumption that similar nodes are more likely to establish an edge in the future. For each unobserved node pair (u, v) in a training network, a link prediction method gives a score $s(u, v)$, which is an estimate of the presence of edge formation between nodes u and v. Let $\Gamma(u)$ denote the neighborhood of node u, i.e. the set of neighbors of node u, and $|\Gamma(u)|$ is the degree of node u. We consider the following similarity measures:

Common neighbor (CN) measures the number of shared neighbors between two nodes. For nodes u and v, the CN is the number of nodes that both u and v have in common. CN is formally defined as:

$$s_{\mathrm{CN}}(u, v) = |\Gamma(u) \cap \Gamma(v)|.$$

Jaccard's coefficient (JC) is a normalized version of CN, which computes the ratio of common neighbors out of all neighbors. Two nodes, u and v, are more likely to have an edge if they have many common neighbors relative to the total number of neighbors they have. An illustration is provided in Figure 8.2. JC is defined as:

$$s_{\mathrm{JC}}(u, v) = \frac{|\Gamma(u) \cap \Gamma(v)|}{|\Gamma(u) \cup \Gamma(v)|}.$$

Adamic/Adar (AA) refines the simple counting of common neighbors by assigning the lower-connected neighbors higher weight. AA is defined as:

$$s_{\mathrm{AA}}(u, v) = \sum_{z \in \Gamma(u) \cap \Gamma(v)} \frac{1}{\log|\Gamma(z)|}.$$

Preferential attachment (PA) PA is the product of the degrees of nodes v and u. Nodes with a higher degree have a stronger ability to grab edges added to the network. PA is formally defined as:

$$s_{\mathrm{PA}}(u, v) = |\Gamma(u)| \cdot |\Gamma(v)|.$$

Based on the derived set of similarity measures, we can also propose supervised (learning-based) link prediction. In supervised settings, each similarity measure serves as a feature vector. In this formulation, link prediction is stated as a binary classification

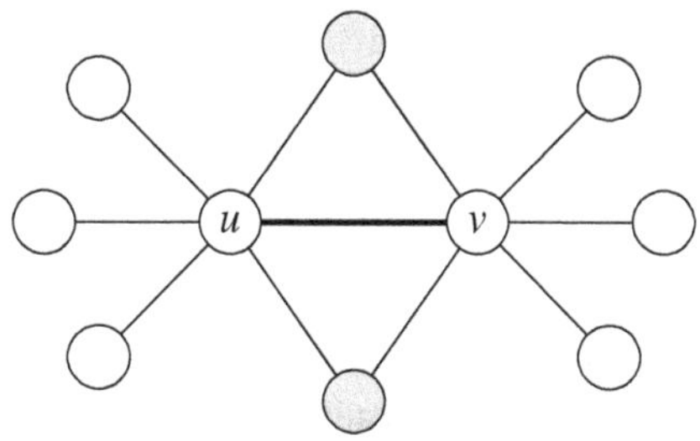

Fig. 8.2 The Jaccard similarity between the sets of neighbors of a pair of nodes measures the number of triangles that would be formed if the two nodes are connected, normalized by the total number of neighbors of the two nodes.

task. Each pair of nodes which is not connected in the training network is treated as a possible instance in the supervised learning setting. If there is an edge connecting this pair of nodes in the training network, this pair is labeled as positive, otherwise, it is labeled as a negative instance. We expect that a classifier that incorporates measures derived under an unsupervised approach can greatly improve predictive performance. We used the following learning algorithms: decision trees (j48), k-nearest neighbors (kNN), logistic regression (LR), multilayer perceptron (MLP), naïve Bayes (NB), and random forests (RF). We refer the reader to any machine learning textbook (e.g., Bishop, 2006) for more detailed information about the classifiers we used.

8.2.3 Performance evaluation of link prediction

We applied the area under the ROC curve (AUC ROC) metric to quantitatively measure the performance of unsupervised and supervised prediction algorithms (see details on ROC analysis in Section 3.7.2). In this study, the ROC analysis approach has been adapted to the link prediction setting as follows.

The link prediction algorithm gives each unobserved edge a score s to quantify its existence in the test dataset. The AUC value can be interpreted as the probability that a randomly chosen missing edge (an edge in E^P) is given a higher score than a randomly chosen non-existing edge (an edge in $U \setminus E$). That is, we randomly select a missing edge and a non-existing edge to compare their scores. If among n independent comparisons, n' times the missing edges have a higher score and n'' times the missing edge and non-existing edge have the same score, the estimator for the AUC value is defined according to Lü and Zhou (2011) as follows:

$$\text{AUC} = \frac{n' + 0.5 \cdot n''}{n}.$$

The degree to which the AUC value exceeds 0.5 indicates how much better the prediction algorithm performs than by chance.

We evaluated the effectiveness of link prediction through a series of experiments using the MeSH network, which contains the history of edge formation between biomedical concepts. We first examine the basic descriptive statistics of the network.

The examined MeSH network consists of 24 401 nodes and 3 464 696 edges. The mean and maximal degree are 284 edges and 7761 edges, respectively. The network's diameter, defined as the longest of the shortest paths between any two nodes, is six edges. The network exhibits the small world property because of a short average path length (2.5 hops) and relatively high clustering (0.45). The giant component of the network comprises about 98% of all nodes. More details about the MeSH network are presented elsewhere (Kastrin et al., 2014a).

We now present the experimental evaluation of the link prediction performance. To assess unsupervised learning, we employed the methodology introduced in Section 8.2.2. In unsupervised settings, link prediction performance for the four similarity measures is summarized in Table 8.1. Each row in the table refers to a five-year network snapshot, which was used as a test set. The corresponding training network was based on edges that co-occur before the given testing period (e.g., for the test network 1996–2000, the training network was built on records from the beginning of MEDLINE to the end of 1995). The AA measure achieved the best performance in terms of AUC, followed by CN, JC, and PA. There was a statistically significant difference between the similarity measures, which was confirmed by the one-way ANOVA test ($p = 0.004$).

Table 8.1 AUC ROC scores for unsupervised link prediction.

Time slice	CN	JC	AA	PA	Mean
1951–1955	0.77	0.64	0.79	0.74	0.74
1956–1960	0.84	0.77	0.85	0.79	0.81
1961–1965	0.74	0.67	0.75	0.71	0.72
1966–1970	0.69	0.69	0.70	0.64	0.68
1971–1975	0.69	0.70	0.71	0.62	0.68
1976–1980	0.63	0.64	0.65	0.54	0.62
1981–1985	0.68	0.69	0.70	0.57	0.66
1986–1990	0.75	0.76	0.77	0.64	0.73
1991–1995	0.75	0.76	0.77	0.64	0.73
1996–2000	0.76	0.77	0.78	0.65	0.74
2001–2005	0.80	0.80	0.82	0.70	0.78
2006–2010	0.82	0.80	0.83	0.73	0.80
Mean	0.74	0.72	0.76	0.66	

Legend: CN = common neighbor, JC = Jaccard coefficient, AA = Adamic/Adar, PA = preferential attachment

Predictive performance results for supervised learning are presented in Table 8.2. We evaluated whether similarity features can be combined to define a model of edge formation across all four predictors. Same as in Table 8.1, the rows of the table designate

Table 8.2 AUC ROC scores for supervised link prediction.

Time slice	j48	kNN	LR	MLP	NB	RF	Mean
1951–1955	0.76	0.72	0.78	0.78	0.72	0.78	0.76
1956–1960	0.85	0.82	0.86	0.86	0.82	0.88	0.85
1961–1965	0.74	0.68	0.76	0.74	0.70	0.73	0.72
1966–1970	0.81	0.77	0.83	0.80	0.70	0.85	0.79
1971–1975	0.84	0.78	0.86	0.83	0.70	0.88	0.82
1976–1980	0.86	0.79	0.87	0.79	0.64	0.89	0.81
1981–1985	0.85	0.81	0.88	0.86	0.69	0.89	0.83
1986–1990	0.87	0.83	0.90	0.89	0.76	0.90	0.86
1991–1995	0.86	0.82	0.90	0.88	0.77	0.90	0.86
1996–2000	0.87	0.82	0.90	0.87	0.77	0.90	0.86
2001–2005	0.88	0.82	0.90	0.89	0.80	0.90	0.86
2006–2010	0.87	0.83	0.88	0.87	0.81	0.90	0.86
Mean	0.84	0.79	0.86	0.84	0.74	0.87	

Legend: j48 = decision trees, kNN = k-nearest neighbors, LR = logistic regression, MLP = multilayer perceptron, NB = naïve Bayes, RF = random forests

five-year windows, which were used for testing. The overall predictive performance is higher than in the unsupervised approach. The RF classifier accomplished the best performance in terms of AUC, followed by LR, j48, MLP, and kNN. There was a statistically significant difference between the performance of classifiers as determined by the one-way ANOVA test ($p < 0.001$).

In summary, the past research on link prediction focused mainly on unsupervised learning. In this study, differences between CN, JC, and AA under an unsupervised approach are low, so no conclusions can be derived about the most appropriate method for link prediction. However, the low performance of the PA measure may indicate the poor fit of the network to a power law distribution as explained in Kastrin et al. (2014b). The results are aligned with Liben-Nowell and Kleinberg (2003) and Zhou et al. (2009), who systematically evaluated local similarity indices on various complex networks.

Empirical evidence showed that CN and AA performed very well, while the PA index had the worst overall performance. On the other hand, supervised learning enables catching significant interdependency relationships in network measures (Lichtenwalter et al., 2010). As expected, supervised learning clearly outperforms unsupervised learning. In supervised settings, the number of positive and negative examples was the same, so in theory, a baseline classifier would have an AUC score of around 0.50. In this study, all supervised classifiers achieve AUC scores around 0.80 and higher. Although the no-free-lunch theorem (Wolpert, 1996) states that no learning method can (always) be the best, we tried to rank the classifiers' performance. The RF classifier performed best; this is not surprising since RF clearly prevails over all state-of-the-art classifiers in a recent exhaustive comparison study of supervised machine learning algorithms

(Fernández-Delgado et al., 2014). Among all classifiers, the performance of NB is the worst. A possible reason for its bad performance could be that a basic assumption of NB was not met: i.e. since three of the four similarity measures used as features are based on observing common neighbors, they are clearly not independent.

8.2.4 Application: Finding connections between schizophrenia and inflammation

As a practical example of our approach, let us investigate the connection between schizophrenia and inflammation. The frequency of MEDLINE records with *Schizophrenia* as a major MeSH term until 2005 is 44 382, while the frequency of records with *Inflammation* is 16 369. If we consider a training network from the beginning of the 20th century to the year 2005 there exists no MEDLINE record that contains both *Schizophrenia* and *Inflammation* as major MeSH terms. The unsupervised approach with Adamic/Adar index predicts a connection between *Schizophrenia* and *Inflammation* in the testing period 2006–2010 with a score of 84.41, the value below which 99.96 % of the score values may be found. The first paper with both MeSH terms appeared in 2006 (Nawa & Takei, 2006). As of this writing, there were 44 records in which both terms co-occurred.

The etiology of schizophrenia remains poorly understood, involving genetic and environmental interactions. By reviewing existing empirical evidence since 2006 it appears that increased inflammation plays a major role in patients with schizophrenia (Kirkpatrick & Miller, 2013). Moreover, a recently conducted meta-analysis supports a growing body of literature that schizophrenia is significantly associated with inflammation (Miller et al., 2014). Researchers have also found that inflammatory response genes have significantly altered expression in schizophrenia (Sainz et al., 2013). Anti-inflammatory strategies may therefore provide an alternative treatment for schizophrenia (Meyer, 2011).

The reader may ask about the prediction validity of our approach. How could someone know that the given example about schizophrenia and inflammation is not a good prediction just by chance? We could address this issue indirectly by examining the topology of the co-occurrence network. It is well known that random networks—in which links occur randomly between the nodes—are typically characterized by Poisson distribution of degree values; otherwise, they exhibit a power law or truncated power law distribution. In our previous exploratory study we showed that a power law provides the best model for data behavior over a considerable portion of the degree distribution of the co-occurrence network considered (Kastrin et al., 2014b). We achieved a very poor fit to the data with Poisson distribution. Moreover, we showed that our network exhibited a hierarchical structure, characterized by clusters of nodes that are connected through a few long-range links.

8.3 Discussion

This chapter presents an approach to LBD using link prediction methods. Link prediction was performed on the MeSH network, a large-scale relational dataset of biomedical concepts. The fundamental conclusions of this study can be summarized as follows. (i) as suggested by the performance evaluation procedure, link prediction proved to be a useful approach for LBD processing using the MeSH vocabulary in the medical domain; (ii) supervised statistical learning approaches outperform unsupervised approaches to link prediction; and (iii) link prediction algorithms are easy to integrate into existing text mining systems.

There are several issues of the current study to be considered. First, the analysis presented relies solely on co-occurrence relations. Co-occurrence represents the simplest way to capture associations between concepts. Co-occurrence, although commonly used, can be interpreted only as an association rather than a substantively meaningful relationship between concepts. Some co-occurrences are also too general to be useful in network modeling (e.g., Humans–Disease). This issue can be overcome by using the SemRep system (Rindflesch & Fiszman, 2003) which introduces relevant semantic relationships between concepts (e.g., Cognitive Therapy TREATS Depressive Symptoms). Using SemRep, the relationships between concepts can be described more precisely and with greater semantic expressiveness. Second, the presented analysis is based only on the MeSH vocabulary. Discovering unknown or unexpected relationships could be especially promising in traditionally independent scientific areas (e.g., psychiatry and herbal medicine). However, different domains, and thereby different vocabularies, could lead to technical challenges. Third, we ignore weights on edges and treat all relationships as equally important. How to properly use weights to boost prediction performance is still an unanswered question (Lü & Zhou, 2011). We expect that the use of weights would significantly improve the prediction performance. Fourth, the analyzed network was originally a sparse network. This means that the set of edges is only a small proportion of all possible edges, which in turn implies an enormous imbalance between positive and negative instances for supervised learning. In our application, we addressed this problem using a simple undersampling strategy. However, it is well known that reducing the majority class has various shortcomings (e.g., can potentially remove certain important examples which may significantly reduce learning performance; Lin & Chen, 2013). In conclusion, analysis of MeSH terms could lead to interesting discoveries, but cannot substitute a deep analysis of the actual contents of the whole papers. Whichever machine analysis we use, the final decision about its usefulness is still in the hands of humans. LBD is exclusively a hypothesis generation tool, which in turn implies that all hypotheses need to be manually evaluated by human experts. In addition, recent initiatives to provide the full content of journal papers offer a great opportunity to improve the implementation of machine-driven link prediction.

8.4 Tutorial in Python

The content presented in this chapter is supplemented by a practical example, for which the source code is hosted on GitHub. Readers are encouraged to access the tutorial **06_link_prediction** at the book's GitHub repository to further explore the concepts discussed. Before executing the provided Jupyter notebook, readers are advised to read Chapter 4 and follow the instructions to install Python and the required packages from **requirements.txt** file included in the repository.

8.4.1 06_link_prediction.ipynb

In this tutorial, we explore how computational techniques can uncover potential new knowledge in the scientific literature by using a co-citation network derived from primary documents. This step-by-step guide introduces an unsupervised approach to link prediction that begins with data preparation and culminates in visualization. We guide the reader through the following phases:

1. *Data preparation*: Acquiring and cleaning co-citation data from the OpenAlex bibliographic database.
2. *Feature extraction*: Extracting relevant proximity measures necessary to predict new links.
3. *Prediction and evaluation*: Using the collection of proximity measures to predict new connections and assessing their accuracy with metrics such as precision and recall.
4. *Visualization and interpretation*: Using visual tools to depict the predicted links, as means to enhance the interpretation of the results.

The dataset includes 1,426 documents from the RS-DFO experiment, expanded into a training network with thousands of cited references for Raynaud's disease and fish oil, obtained from the OpenAlex database. The testing network includes the first 1,000 cited references published after November 30, 1985. The notebook addresses the unsupervised link prediction using measures such as common neighbors, the Jaccard coefficient, and the Adamic/Adar index. A time-sliced evaluation confirms the high accuracy of this method, underscoring the effectiveness of the link prediction approach for LBD.

References

Alako, B. T. F., Veldhoven, A., van Baal, S., Jelier, R., Verhoeven, S., Rullmann, T., Polman, J., & Jenster, G. (2005). CoPub Mapper: Mining MEDLINE based on search term co-publication. *BMC Bioinformatics*, *6*, 51.

Bales, M. E., & Johnson, S. B. (2006). Graph theoretic modeling of large-scale semantic networks. *Journal of Biomedical Informatics*, *39*(4), 451–464.

Bishop, C. M. (2006). *Pattern Recognition and Machine Learning*. Springer.

Cohen, K. B., & Hunter, L. (2008). Getting started in text mining. *PLoS Computational Biology*, *4*(1), 0001–0003. https://doi.org/10.1371/journal.pcbi.0040020

Fernández-Delgado, M., Cernadas, E., Barro, S., & Amorim, D. (2014). Do we need hundreds of classifiers to solve real world classification problems? *The Journal of Machine Learning Research*, *15*(1), 3133–3181.

Hoffmann, R., & Valencia, A. (2005). Implementing the iHOP concept for navigation of biomedical literature. *Bioinformatics*, ii252–258.

Kastrin, A., Rindflesch, T. C., & Hristovski, D. (2014a). Large-scale structure of a network of co-occurring MeSH terms: Statistical analysis of macroscopic properties. *PloS One*, *9*(7), e102188.

Kastrin, A., Rindflesch, T. C., & Hristovski, D. (2014b). Large-scale structure of a network of co-occurring MeSH terms: Statistical analysis of macroscopic properties. *PloS one*, *9*(7), e102188.

Kastrin, A., Rindflesch, T. C., & Hristovski, D. (2014c). Link prediction on the semantic MEDLINE network. *Proceedings of the 17th International Conference on Discovery Science*, 135–143.

Kirkpatrick, B., & Miller, B. J. (2013). Inflammation and schizophrenia. *Schizophrenia Bulletin*, *39*(6), 1174–1179.

Kumar, A., Singh, S. S., Singh, K., & Biswas, B. (2020). Link prediction techniques, applications, and performance: A survey. *Physica A: Statistical Mechanics and its Applications*, *553*, 124289.

Li, C., Liakata, M., & Rebholz-Schuhmann, D. (2013). Biological network extraction from scientific literature: State of the art and challenges. *Briefings in Bioinformatics*, *15*(5), 856–877.

Liben-Nowell, D., & Kleinberg, J. (2003). The link prediction problem for social networks. *Proceedings of the 12th International Conference on Information and Knowledge Management*, 556–559.

Lichtenwalter, R. N., Lussier, J. T., & Chawla, N. V. (2010). New perspectives and methods in link prediction. *Proceedings of the 16th ACM SIGKDD international conference on Knowledge discovery and data mining*, 243–252.

Lin, W.-J., & Chen, J. J. (2013). Class-imbalanced classifiers for high-dimensional data. *Briefings in Bioinformatics*, *14*(1), 13–26.

Lü, L., & Zhou, T. (2011). Link prediction in complex networks: A survey. *Physica A: statistical mechanics and its applications*, *390*(6), 1150–1170.

Martinez, V., Berzal, F., & Cubero, J.-C. (2016). A survey of link prediction in complex networks. *ACM Computing Surveys*, *49*(4), 1–33.

Meyer, U. (2011). Anti-inflammatory signaling in schizophrenia. *Brain, Behavior, and Immunity*, *25*(8), 1507–1518.

Miller, B. J., Culpepper, N., & Rapaport, M. H. (2014). C-reactive protein levels in schizophrenia: A review and meta-analysis. *Clinical Schizophrenia & Related Psychoses*, *7*(4), 223–230.

Nawa, H., & Takei, N. (2006). Recent progress in animal modeling of immune inflammatory processes in schizophrenia: Implication of specific cytokines. *Neuroscience Research*, *56*(1), 2–13.

Rindflesch, T. C., & Fiszman, M. (2003). The interaction of domain knowledge and linguistic structure in natural language processing: Interpreting hypernymic propositions in biomedical text. *Journal of Biomedical Informatics*, *36*(6), 462–477.

Sainz, J., Mata, I., Barrera, J., Perez-Iglesias, R., Varela, I., Arranz, M. J., Rodriguez, M. C., & Crespo-Facorro, B. (2013). Inflammatory and immune response genes have significantly altered expression in schizophrenia. *Molecular Psychiatry*, *18*(10), 1056–1057.

Thomas, P., Starlinger, J., Vowinkel, A., Arzt, S., & Leser, U. (2012). GeneView: A comprehensive semantic search engine for PubMed. *Nucleic Acids Research*, *40*(Web Server issue), W585–591.

Wolpert, D. H. (1996). The lack of a priori distinctions between learning algorithms. *Neural Computation*, *8*(7), 1341–1390.

Yu, Q., Long, C., Lv, Y., Shao, H., He, P., & Duan, Z. (2014). Predicting Co-Author Relationship in Medical Co-Authorship Networks. *PLOS ONE*, *9*(7), e101214.

Zhou, T., Lü, L., & Zhang, Y.-C. (2009). Predicting missing links via local information. *The European Physical Journal B*, *71*(4), 623–630.

Chapter 9
Embedding-based Closed Discovery

This chapter presents embedding-based approaches to bisociative literature-based discovery.[1] Section 9.1 explains the concept of embeddings and illustrates the potential of word embeddings for bisociative knowledge discovery. Section 9.2 presents a text embedding-based LBD methodology, and a network embedding-based LBD approach presented in Section 9.3. An implementation of the text embedding-based approach to LBD is presented in Section 9.4.

9.1 Embeddings

Regarding representation learning, our text-based LBD approaches (Cestnik et al., 2017; Juršič et al., 2012; Petrič et al., 2012; Sluban et al., 2012) (outlined in Chapter 5 and Chapter 6) and other early LBD approaches used the standard TF-IDF (Term Frequency Inverse Document Frequency) Bag-of-Words (BoW) vector representations of text documents (see Section 3.3). On the other hand, LBD approaches can also be based on embedding-based representations (see Section 3.5).

The learning task of automated embedding construction is defined as follows.

Learning embeddings. **Given** input data of a given data type and format, **find** a tabular representation of the data, where each row represents a single data instance, and each column represents one of the dimensions in the d-dimensional numeric vector space $\mathbb{R}^d$.

In one-hot encoding, each word is represented by a vector where all components are set to zero, except for one element that is assigned a value of 1, corresponding to the specific word. However, in such a representation, the meaning of words is not preserved by the word vectors themselves.

Word embedding, in its simplest form, is a mapping from a vocabulary to a vector space. Its origins lie in the concept of distributed representations (Hinton et al., 1986), where the meaning of a word is spread across multiple features. Figure 9.1 illustrates

[1] This chapter is largely based on the work of Lavrač et al. (2020) and Zhang et al. (2021).

N. Lavrač et al., *Bisociative Literature-Based Discovery*,
https://doi.org/10.1007/978-3-031-96863-1_9

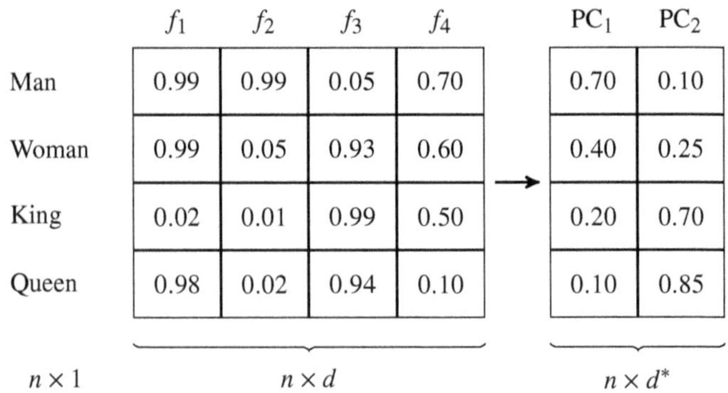

Fig. 9.1 Word embedding workflow. The figure illustrates the transition from a vocabulary vector to an embedding matrix followed by PCA for visualization. See text for details.

three key components of a typical word embedding workflow: the vocabulary prepared from a set of documents is represented as a column vector of length n; an embedding matrix of size $n \times d$, where each row corresponds to a word's embedding vector learned via the word2vec algorithm; and a matrix of size $n \times 2$, resulting from applying PCA on the embedding matrix to facilitate 2D visualization while preserving structural relationships between words. For example, words such as "king" and "queen" might share a common "royalty" feature. This approach is based on the distributional hypothesis, which suggests that words can be understood through the contexts in which they appear and the words by which they can be replaced. By analyzing these contextual relationships, word embeddings capture and represent the meanings of words in a latent, high-dimensional space.

Embeddings are usually obtained by training algorithms such as word2vec (Mikolov, Chen, et al., 2013), GloVe (Pennington et al., 2014), or FastText (Bojanowski et al., 2017), which characterize the word based on the semantic or lexical context in which it appears. These representations improve performance on a variety of automated text processing tasks because they can capture the semantic meaning of words, because words with similar meanings are represented by vectors that are close together in the embedding space. This property also enables the capture of regularities that go beyond simple relatedness, including analogies (Mikolov, Le, & Sutskever, 2013).

A well-known example illustrating this notion is that word embeddings can explicitly find relationships between words and discover analogies between pairs of words, such as that the relationship between *man* and *woman* is very similar to that between *uncle* and *aunt*, or to the relationship between *king* and *queen*, shown in Figure 9.2.

Embedding technology can be used to discover analogies within a single domain, as shown in Figure 9.2. In addition, research on cross-lingual embeddings (Conneau et al., 2017) has shown the ability of the embeddings technology to match embedding spaces across languages, which can be used as a basis for finding analogies across text corpora in different languages or contexts.

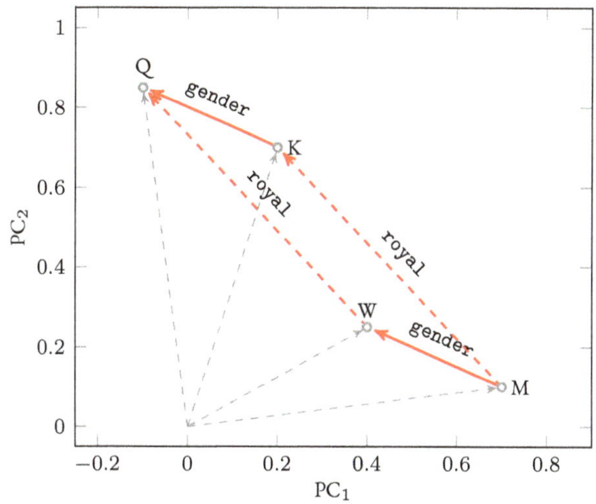

Fig. 9.2 Two-dimensional projection of word embeddings, illustrating the relations between words man (M), woman (W), king (K) and queen (Q). Word embedding vectors are shown with gray dashed lines, while the relations between words are depicted in red. Notice that embeddings allow for detecting analogies, e.g., the gender relationship between *man* and *woman* being very similar to the gender relationship between *king* and *queen*.

9.2 Text embedding-based bisociative LBD methodology

For bisociative LBD, the most important property of word embeddings is that they can capture regularities beyond simple relatedness, such as analogies (Mikolov, Le, & Sutskever, 2013) illustrated in Figure 9.2. This section demonstrates that, in analogy to cross-lingual embeddings, word embeddings can be effectively used also in the context of cross-domain bisociative literature-based discovery. The presented method is based on the idea of mapping the cross-lingual embeddings to a cross-domain setting: instead of considering two different languages, we consider two separate domains A and C, apply embedding alignment methods (Conneau et al., 2017) to align related concepts in the two domains, and finally perform analogy detection across the two domains. This approach allows us to implement the idea of *bridging by relational bisociations*, which is illustrated in Figure 9.3.[2]

In the closed literature-based discovery setting of our interest, we implement the concept of bridging by relational bisociation as follows.

Bridging by relational bisociation. In the *bridging by relational bisociation* setting, illustrated in Figure 9.3, we investigate whether, for a specific relation between two given concepts $a1$ and $a2$ in the first domain A, there exists an analogous relation between concepts x and c in the second domain C, where c is a given concept and

[2] Note that the concept of bridging by relational bisociations was previously mentioned in Section 1.3.5 and illustrated in Figure 1.9.

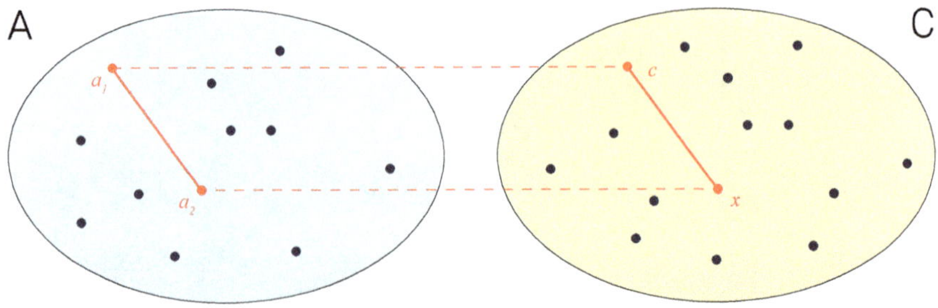

Fig. 9.3 Concept of bridging by relational bisociation.

x is a new concept yet to be discovered. More formally, a bisociation between two separate domains A and C is defined as an analogy between:

```
a1 rel a2 == c rel x.
```

In the embedding space, this analogy translates to the following equation between embeddings (vectors in an n-dimensional vector space):

```
emb(x) = emb(a2) - emb(a1) + emb(c).
```

Finally, once x is calculated, we need to find a set of concepts from the second domain C that have an embedding representation most similar to x according to some predefined distance measure. For various distance measures, including the cosine, Euclidean and Manhattan distances, see Section 3.3.2.

9.2.1 Methodology of bridging by relational bisociation

The embedding-based bisociative LBD methodology for discovering bisociated relationships between two domains A and C, proposed by Lavrač et al. (2020), consists of the following steps:

1. Select two domains A and C.
2. Train separate word embedding models for A and C to get $emb(A)$ and $emb(C)$.
3. Align the embedding vector spaces $emb(A)$ and $emb(C)$.
4. Determine the relationships of interest in a given domain A between concepts $a1$ and $a2$ defined by the domain expert.
5. Perform embedding-based relational LBD with a known seed concept c in C by leveraging the ability of embedding representations to model analogy relations.
6. Evaluate the list of best-ranked relational bisociations.

9.2.2 Application: Finding connections between circadian rhythm and plant defense

This section presents LBD study conducted on two biology literature domains, the *circadian rhythm* and *plant defense* domains, where the goal was to identify novel daily regulated mechanisms that are responsible for the plant defense. The circadian rhythm in plants causes some of the plant genes to be expressed differently during the day. Consequently, plants respond differently to disease-causing infection if they are infected at different times of the day (e.g., morning, noon, evening). Therefore, one of the goals of our study was to identify new sets of genes that are expressed differently in different parts of the day and are important for the defense of plants against the pathogen.

After obtaining 10,494 PubMed documents, containing article titles and abstracts (4,346 from plant defense and 6,148 from circadian rhythm domains), we replaced gene names with synonyms that were gathered in previous research projects (22,265 gene names mapped into 7,863 synonyms). In addition, we pre-processed the documents to keep only gene-related terms (included in the synonyms list and from the gene dictionary containing 6,083 additional gene names), which resulted in a substantially reduced document corpus. The experiments carried out by the methodology proposed in Section 9.2 served as a proof of concept that the proposed embedding-based methodology can be used for LBD.

For each of the two selected domains, *circadian rhythm* and *plant defense*, we trained a separate FastText embedding model (Bojanowski et al., 2017). FastText embeddings were chosen due to their ability to leverage both semantic and morphological information by representing each word as an average of its character n-grams. This is useful in a setting with relatively small domain corpora containing limited semantic information since morphological similarity in many cases translates to semantic relatedness.

The resulting embedding models trained for each domain were aligned in the next step in a common vector space. We selected a supervised alignment approach, which is based on a training dictionary of identical words from both domains that are used as anchor points to learn a mapping from the source to the target space with a Procrustes alignment (Conneau et al., 2017). Train and test dictionaries were constructed by taking 5,000 most frequent words from both domains (i.e. the words that appear in both domains and have the largest sum of frequencies) and then randomly split into a training dictionary containing two-thirds of the words (3,333) and a test dictionary containing one-third of the words (1,667).

The success of the alignment was measured on the test dictionary in terms of precison@k, where precison@1 represents a share of the model's correct alignments (exact matches) in a set of all alignments, and precison@5 represents a share of the model's alignments in a set of all alignments, where the correct match for the word is found in the set of five most probable alignments predicted by the model. In the experiment, we report the precison@1 of 0.4 and precison@5 of 0.55.

Next, we asked the biology expert to identify a list of genes related to the circadian rhythm domain. The following list was produced:

1. CCA1 = CIRCADIAN CLOCK ASSOCIATED1

2. LHY = LATE ELONGATED HYPOCOTYL
3. TOC1 = TIMING OF CAB EXPRESSION 1
4. PRR1 = PSEUDO-RESPONSE REGULATOR 1
5. GI = GIGANTEA
6. LNK1 = NIGHT LIGHT-INDUCIBLE AND CLOCK-REGULATED 1
7. PRR5 = PSEUDO-RESPONSE REGULATOR 5
8. ELF4 = EARLY FLOWERING 4
9. PRR9 = PSEUDO-RESPONSE REGULATOR 9
10. PRR7 = PSEUDO-RESPONSE REGULATOR 7
11. PCL1 = PHYTOCLOCK 1
12. ELF3 = EARLY FLOWERING 3

In addition, we also took more general key concepts from the circadian rhythm domain:

13. NEGATIVE FEEDBACK LOOP
14. OSCILLATOR
15. CLOCK

Following the methodology explained in Section 9.2, we tried to identify a list of genes related to the concept of *plant defense*, which are related to the concept of *circadian rhythm* in a similar way as the expert-listed genes are related to the circadian rhythm concept. First, we calculated the embedding x according to the following equation:

```
emb(x) = emb(a2) - emb(a1) + emb(c).
```

where `a1` is the concept *circadian rhythm*, `a2` is a gene from the above list and `c` is the concept *plant defense*.

Finally, once `x` was calculated for each of the genes in the above list, we searched for a set of concepts of the plant defense domain that have an embedding representation most similar to `x` according to the cosine similarity measure (see Equation 3.2 in Section 3.3.2). To limit the results only to genes or gene-related concepts, the concepts from the second domain were considered only if they appeared in the reduced document corpus. 10 genes or gene-related concepts with the representation most similar to each of the calculated `x`'s were identified and given to the biology expert for evaluation.

9.2.3 Results evaluation

The domain expert evaluated the selected set of output terms for all given analogy inputs. More specifically, for the analogies—*a2 is as important to a1 (circadian rhythm) as x is to c (plant defense)*—for each input relation, the resulting list of 10 candidates most similar to x (according to the cosine similarity between the candidate's embedding and x) were evaluated by the expert who was given instructions to manually classify the relatedness between a candidate and the plant defense domain into the following four categories: NOT AT ALL; NOT REALLY; MAYBE; YES. While YES is the category serving as proof that the methodology works, MAYBE is the category containing very interesting terms from the knowledge discovery point of view, as here the expert may be stimulated to search for novel knowledge.

Table 9.1 Evaluation for 15 input relations.

Source gene/term	Not at all	Not really	Maybe	Yes	P@10 Maybe	P@10 Yes
CCA1	1	3	1	5	0.1	0.5
LHY	0	5	1	4	0.1	0.4
TOC1	0	1	3	6	0.3	0.6
PRR1	0	5	2	3	0.2	0.3
GI	1	5	2	2	0.2	0.2
LNK1	1	4	1	4	0.1	0.4
PRR5	0	3	3	4	0.3	0.4
ELF4	0	2	2	6	0.2	0.6
PRR9	0	4	0	6	0.0	0.6
PRR7	0	3	1	6	0.1	0.6
PCL1	2	6	0	2	0.0	0.2
ELF3	1	4	2	3	0.2	0.3
NEGATIVE FEEDBACK LOOP	7	0	2	1	0.2	0.1
OSCILLATOR	4	2	1	3	0.1	0.3
CLOCK	1	1	3	5	0.3	0.5
All	18	48	24	60	0.16	0.4

First, we calculated the average precision at 10 (p@10) for each output list of 10 candidates, as well as the micro-averaged precision for the entire dataset (see Table 9.1). We can observe that the method performed very well. In 40% of the cases, the expert found in the scientific literature that the discovered relation between the plant defense concept and the proposed term **x** is meaningful. We can see that precision varies for different input relations, but the method was able to find at least one correct relation in the plant defense domain for each circadian rhythm input relation. For input relations between the concept *circadian rhythm* and genes ELF4, PRR9 and PRR7, six out of ten term candidates in the resulting candidate lists are related to the plant defense domain. On the other hand, the lowest results are for the input concept *negative feedback loop*, where only for one out of ten output terms the expert found that the output term was relevant for the domain. A reason for this could be that the input term is one of the few terms that is not a gene but rather a gene related concept (text), and that it is a multi-word expression, for which the average embedding was first calculated (by averaging embeddings for each word in the term) to obtain the term embedding, and therefore the results might be less precise.

For the category MAYBE, which is the most interesting category for new knowledge discovery, we can note that for all input relations but one, at least one out of 10 outputs was considered potentially interesting. In a knowledge discovery setting, this is considered a promising result.

As explained above, the biology expert evaluated 10 candidates for each input relation. These relations were ranked according to the cosine similarity between x and the candidate, with rank 1 representing the candidate closest to x, i.e. with the largest cosine similarity to x. Table 9.2 presents the results for candidates with different ranks. Note that here we measured precision at 15, i.e. how many out of 15 predicted terms with a specific rank had been evaluated as related to the plant defense domain (P@15 YES) or as maybe being related to the plant defense domain (P@15 MAYBE). Interestingly,

Table 9.2 Evaluation according to rank.

Rank	Not at all	Not really	Maybe	Yes	P@15 Maybe	P@15 Yes
1.	2	4	1	8	0.067	0.533
2.	2	6	1	6	0.067	0.400
3.	1	7	1	6	0.067	0.400
4.	3	7	2	3	0.133	0.200
5.	1	3	2	9	0.133	0.600
6.	4	1	3	7	0.200	0.467
7.	1	4	4	6	0.267	0.400
8.	2	3	5	5	0.333	0.333
9.	1	6	3	5	0.200	0.333
10.	1	7	2	5	0.133	0.333
All	18	48	24	60	0.160	0.400

the correlation between precision at 15 and rank was not strong, and the better-ranked candidates were not necessarily more correlated to the plant defense domain according to the evaluation. For example, the best-evaluated candidates had rank 5, where P@15 YES was 0.6 and P@15 MAYBE was 0.133.

Table 9.3 Evaluation on all output terms (duplicates removed).

Label	Count	Percentage
NOT AT ALL	14	19.18%
NOT REALLY	19	26.03%
MAYBE	13	17.81%
YES	27	36.99%
Total	73	100%

Next, we removed duplicate outputs, merged all output terms from all inputs, and calculated the class distribution for this list (see Table 9.3). The rationale for this procedure is that since in our case a1 and c were always the same (equivalent to circadian rhythm and plant defense domain names) and since the different a2 all modeled the same relation—*a2 is as important to a1 (circadian rhythm) as x is important to c (plant defense)*—we could also treat all results as a common list of relevant terms (genes). As the results indicate, about 37% of the output terms were evaluated as relevant to the plant defense domain. Also, together with the category MAYBE, which indicates that the output is potentially relevant (but requires further research), this percentage of relevant terms increased to nearly 55%.

From 40 examples in categories YES and MAYBE, 32 were gene names, 3 were proteins, while the rest referred to a disease or partial names of proteins, genes etc. Below we list ten terms and their full names that were classified in category YES and appeared in the results of at least 3 input terms:

1. DMR1 = DOWNY MILDEW RESISTANT 1
2. CPR30 = CONSTITUTIVE EXPRESSER OF PR GENES 1
3. EIF4G = EUKARYOTIC TRANSLATION INITIATION FACTOR 4 G
4. SLAC1 = SLOW ANION CHANNEL-ASSOCIATED 1
5. RFC3 = REPLICATION FACTOR C SUBUNIT 3

6. RTM1 = RESTRICTED TEV MOVEMENT 1
7. SNI1 = SUPPRESSOR OF NPR1-1
8. GRF6 = GROWTH-REGULATING FACTOR 6
9. NAC083 = NAC DOMAIN CONTAINING PROTEIN 83
10. XAP5 = XAP5 CIRCADIAN TIMEKEEPER

In summary, the outlined methodology of using advanced embedding technology for bisociative cross-domain literature mining proved to be useful for finding previously unrelated concepts in the closed discovery setting. One of the most interesting findings was the presence of some resistance and susceptibility genes among the candidates proposed by the method; these genes are known to play an important role in plant defense. The best-ranked candidate obtained for the *c* term inputs CCA1 and LHY (two central genes of the circadian clock rhythm) was DMR1 (a susceptibility gene, mutation of this gene results in a higher resistance), which is a hot topic in recent plant resistance research.

9.3 Network embedding-based bisociative LBD methodology

Computational approaches to drug repurposing have garnered much attention, for example, to speed up the discovery of therapies for COVID-19 (Altay et al., 2020; Wang & Guan, 2020). Common computational drug repurposing methods include drug signature matching, molecular docking, genome-wide association studies, and network analysis (Pushpakom et al., 2019). These data-driven approaches involve systematic analysis of various types of biological and clinical data (e.g., gene expression, chemical structure, genome and protein sequences, and electronic health records) to generate hypotheses regarding the repurposing of approved or investigational drugs (Pushpakom et al., 2019). The potential of recent advances in machine learning for COVID-19 drug repurposing has also been demonstrated (Zhou, Wang, et al., 2020), including approaches that leverage network medicine principles and biological knowledge graphs.

Several studies using advanced machine learning techniques have reported promising results (e.g., Zhou, Hou, et al., 2020). Most computational approaches to drug repurposing have focused on biological data, such as gene expression, protein-protein and drug-target interactions, and SARS-CoV-2-related data. However, COVID-19-specific data are meaningful in the context of a larger body of diverse knowledge in medicine and life sciences, the primary source of which is the biomedical literature. Several COVID-19 drug repurposing studies incorporated literature-based knowledge (Ge et al., 2020; Zeng et al., 2020), however, their focus has remained largely COVID-19 specific. We argue that approaches to efficiently and safely repurposing drugs to treat COVID-19 should be based on a more effective integration of literature knowledge with biological data collected with high-throughput methods.

This section presents an effective LBD approach to COVID-19 drug repurposing. Unlike most approaches that focus on COVID-19-specific knowledge (e.g., Ge et al., 2020), we considered a larger body of biomedical knowledge captured in the PubMed bibliographic database, and took a subset of semantic relations derived from the PubMed dataset. In this work, we cast drug repurposing as a knowledge graph completion task to answer the question of how to accurately predict future co-citation links between research papers in previously disconnected research domains, by using state-of-the-art neural network-based algorithms (Bordes et al., 2013; Sun et al., 2019; Trouillon et al., 2016; Yang et al., 2014).

9.3.1 Knowledge graph preprocessing

Our data source was a literature-derived biomedical knowledge graph, constructed from SemMedDB (Kilicoglu et al., 2012), assumed accurate and informative for drug repurposing.

First, we eliminated relations involving generic biomedical concepts (i.e. relations in which both subject and object were present in the `GENERIC_CONCEPT` table of SemMedDB such as `Pharmaceutical Preparations`) and relations with identical subject and object arguments. Next, we excluded a subset of predicate types that were not expected to be useful for drug repurposing, such as PART_OF and PROCESS_OF. The predicate types we used are AFFECTS, ASSOCIATED_WITH, AUGMENTS, CAUSES, COEXISTS_WITH, COMPLICATES, DISRUPTS, INHIBITS, INTERACTS_WITH, MANIFESTATION_OF, PREDISPOSES, PREVENTS, PRODUCES, STIMULATES, and TREATS. Lastly, we also excluded the relations in which the subject or the object belongs to one of the following semantic groups: Activities & Behaviors, Concepts & Ideas, Objects, Occupations, Organizations, and Phenomena. After elimination, the knowledge graph consisted of 331,427 unique nodes and 20,017,236 relations.

In the second step, we eliminated (i) high-degree concepts using network degree centrality, and (ii) uninformative semantic relations using the log-likelihood ratio. The adjacency matrix A of a knowledge graph (i.e. directed network with multi-edges) with n nodes (i.e. concepts) has entries $A_{ij} = 1$ if there is a relation from concept i to concept j. Following Boccaletti et al. (2006), the in- and out-degrees of concept i can then be expressed as :

$$k_i^{\text{in}} = \sum_{j=1}^{n} A_{ji}, \quad \text{and} \quad k_i^{\text{out}} = \sum_{j=1}^{n} A_{ij}.$$

To filter out uninformative links, we assigned each semantic relation a G^2 score, defined in Eq. 9.1 below, indicating how strongly the terms within a triple are associated with each other (McInnes, 2004). High G^2 score means that the observed and expected frequencies are significantly different, indicating that the triple is less likely to occur by chance. For computational purposes, we created two three-dimensional contingency tables with indices i, j, and k. The first table (OT) holds observed frequencies of a triple from the knowledge graph and the second table (ET) contains the expected values, assuming the independence of terms in each triple. G^2 was then calculated using the equation

$$G^2 = 2 \cdot \sum_{i,j,k} n_{ijk} \cdot \log\left(\frac{n_{ijk}}{m_{ijk}}\right), \quad m_{ijk} = \frac{\sum_i n_{jk} \cdot \sum_j n_{ik} \cdot \sum_k n_{ij}}{T^2}, \tag{9.1}$$

where n_{ijk} is the cell i, j, k in OT, m_{ijk} is the cell i, j, k in ET, and $T = \sum n_{ijk}$.

Next, all three measures (G^2, k_i^{in}, and k_i^{out}) were normalized to the range $[0, 1]$ and summed up into a final score. The lower the score, the more specific and informative the relation is. For example, the relation `Operative Surgical Procedures`-TREATS-`Woman`, which has a high score, is more general than the relation `interleukin-6`-AFFECTS-`Autoimmune Diseases`. We also kept all relations with biomedical concepts that refer to the Covid-19 terms in UMLS:[3]

- `C5203670:COVID19 (disease)`
- `C5203671:Suspected COVID-19`
- `C5203672:SARS-CoV-2 vaccination`
- `C5203673:Antigen of SARS-CoV-2`
- `C5203674:Antibody to SARS-CoV-2`
- `C5203675:Exposure to SARS-CoV-2`
- `C5203676:SARS-CoV-2`

We estimated that approximately 2.5M relations could be processed in a reasonable amount of time with our GPU and eliminated relations with high final scores. At the end of the preprocessing stage, the knowledge graph consisted of 131,355 nodes and 2,558,935 relations.

9.3.2 Knowledge graph completion

Consider a knowledge graph $\mathcal{G} = (\mathcal{E}, \mathcal{R}, \mathcal{E})$, where $\mathcal{E}$ refers to a set of entities, $\mathcal{R}$ denotes a set of possible relations, and $\mathcal{T}$ stands for a set of triples in the form (h)ead-(r)elation-(t)ail, formally denoted as $\{(h, r, t)\} \subset \mathcal{E} \cdot \mathcal{R} \cdot \mathcal{E}$. The aim of knowledge graph completion is to infer new triples (h', r', t') such that $h', t' \in \mathcal{E}$ and $r' \in \mathcal{R}$. In this setting, the knowledge graph completion problem can be represented as a ranking task in which a prediction function $\psi(h, r, t) : \mathcal{E} \cdot \mathcal{R} \cdot \mathcal{E} \mapsto \mathbb{R}$, which generates higher scores for true triples and lower scores for false triples is learned.

We used state-of-the-art neural network-based algorithms (Bordes et al., 2013; Sun et al., 2019; Trouillon et al., 2016; Yang et al., 2014), belonging to two classes of knowledge graph completion methods: (i) *translational* models: TransE (Bordes et al., 2013) and RotatE (Sun et al., 2019), and (ii) *semantic matching* models: DistMult (Yang et al., 2014) and ComplEx (Trouillon et al., 2016). These methods differ in the way that they encode entities and relations in a knowledge graph into a low-dimensional vector space (i.e. knowledge graph embedding). Such distributed vector representations can be used for downstream reasoning and machine learning.

- TransE embeds entities and relations into a low-dimensional vector space, learning the embedding vectors for entities and relations from a training set of triplets (h, r, t), composed of head h and tail t entities from entity set E, and a relation r from relation set R, such that $\mathbf{h} + \mathbf{r} \approx \mathbf{t}$, where $\mathbf{h}, \mathbf{r}, \mathbf{t} \in \mathbb{R}^d$ are the embeddings of h, r, and t, respectively. To measure the plausibility of the relations, TransE employs

[3] https://lhncbc.nlm.nih.gov/ii/tools/MetaMap/run-locally/Covid19Terms.html

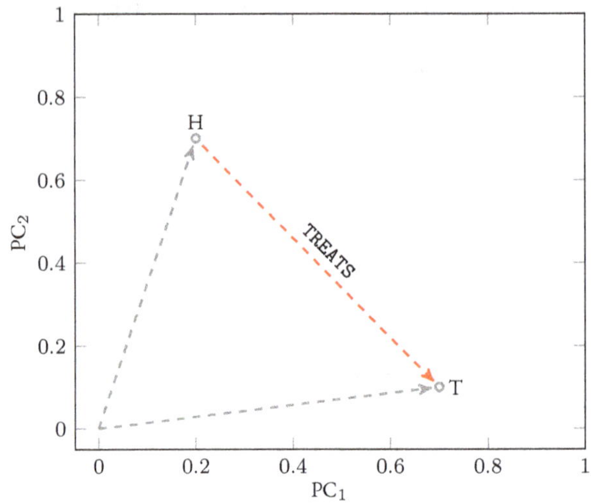

Fig. 9.4 TransE models relations as translations on a low-dimensional embedding of the entities. The head and tail entities are represented by the vector points H and T in the TransE embedding space. The embedding of the tail entity t (i.e., `Covid-19`) should be close to the embedding of the head entity h (i.e., `Metoclopramide`) plus vector that depends on the relationship r (i.e., TREATS).

a distance-based scoring function $s(h, r, t) = ||\mathbf{h} + \mathbf{r} - \mathbf{t}||$, using the norms L_1 or L_2. Figure 9.4 illustrates the TransE model with two-dimensional embeddings. We chose TransE because of its simplicity and good prediction performance. However, TransE can model only one-to-one relations and fails to embed one-to-many, many-to-one, and many-to-many relations. To solve these problems, several solutions were proposed, including RotatE.

- RotatE treats each relation in a complex vector space as a rotation from the head entity to the tail entity, i.e. $s(h, r, t) = |\mathbf{h} \circ \mathbf{r} - \mathbf{t}|_{l_1}$, where $\circ$ is a Hadamard product. We used RotatE instead of TransE, as TransE reportedly does not perform well on some data sets (e.g., FB15k benchmark data set (Bordes et al., 2013), commonly used in knowledge graph completion performance studies), which require symmetric pattern modeling.
- DistMult is the simplest approach among semantic matching models. Its scoring function is defined as $s(h, r, t) = \langle \mathbf{h}, \mathbf{r}, \mathbf{t} \rangle$. However, DistMult is limited only to symmetric relations, generating the same scores for triples (h, r, t) and (t, r, h).
- ComplEx extends DistMult to the complex domain. The head and tail embeddings for the same entity are complex conjugates, enabling ComplEx to model asymmetric relations. Its score function is defined as $s(h, r, t) = \mathrm{Re}(\mathbf{h} \odot \mathbf{r} \odot \mathbf{t})$, where $\mathbf{h}, \mathbf{r}, \mathbf{t} \in \mathbb{C}^k$, $\odot$ denotes the Hadamard product, $\mathrm{Re}(\cdot)$ extracts the real part of the complex vector, and k is the embedding dimension.

The hyperparameters for both sets of models were tuned using the grid search on a validation set for each prediction model. We tuned the learning rate $\eta \in \{0.001, 0.01, 0.1\}$, number of hidden dimensions $k \in \{50, 100, 250, 400\}$, regularization coefficient

Table 9.4: Knowledge graph completion performance.

Model	MR	MRR	Hits@1	Hits@3	Hits@10
TransE	9.223	0.525	0.417	0.585	0.699
DistMult	11.639	0.325	0.216	0.340	0.515
ComplEx	11.045	0.332	0.216	0.352	0.553
RotatE	10.864	0.377	0.246	0.428	0.633

Notes: MR = mean rank, MRR = mean reciprocal rank, Hits@k = fraction of true entities that appear in the first k entities of the sorted rank list.

$\lambda \in \{2 \cdot 10^{-6}, 2 \cdot 10^{-8}\}$, negative adversarial sampling $\in$ {`True`, `False`}, fixed margin $\gamma \in \{1, 5, 10, 20\}$ for RotatE and norm $d \in \{L_1, L_2\}$ for TransE model.

The knowledge graph completion results for all employed models are presented in Table 9.4 in terms of mean rank (MR), mean reciprocal rank (MRR), and Hits@k (for $k \in \{1, 3, 10\}$), as defined in Section 3.7.3 of Chapter 2. For MR, a lower score is considered better; for all others, a higher score is considered better. The score for each method is the mean value over all triplets in the testing set.

On average, TransE outperforms all counterparts on all performance measures. Optimal TransE configuration was achieved with $k = 400$ hidden dimensions, L_1 norm, $\eta = 0.01$ learning rate, and $\lambda = 2 \times 10^{-8}$ regularization coefficient. Model training was limited to 20,000 epochs. Relatively small number of relations (15) ensure that all entities and relations can be smoothly embedded into the same vector space.

9.4 Tutorials in Python

The content presented in this chapter is supplemented by two practical examples whose source code is hosted on GitHub. We encourage readers to access the tutorials **07_find_genes** and **08_drug_repurposing** in the book's GitHub repository to further explore the concepts covered here. Before running the provided Jupyter notebook, the readers are advised to read Chapter 4 and follow the instructions for installing Python and the required packages from the **requirements.txt** file included in the repository.

9.4.1 07_find_genes.ipynb

The workflow in the 07_find_genes.ipynb notebook consists of the following main steps: setting up the environment and importing libraries, generating datasets, embedding learning and alignment, and identifying similar gene relationships in two domains. In the last step, the most similar relationships between genes in the plant defense domain are analyzed for pre-selected relationships between genes in the circadian rhythm domain (e.g. CCA1 rel. PRR7). For each preselected relationship in the circadian rhythm domain, ten of the most similar relationships between genes in the plant defense domain are listed.

9.4.2 08_drug_repurposing.ipynb

The tutorial on drug repurposing is designed to build towards predicting and evaluating potential new uses of existing drugs for Covid-19. The workflow in the `08_drug_repurposing.ipynb` notebook consists of the following steps:

1. Data acquisition: This first phase focuses on collecting a comprehensive subset of semantic relations derived from the SemMedDB datasets, predicted to be accurate and informative for drug repurposing.
2. Data preprocessing: After data collection, the next step involves cleaning and standardizing the acquired data to ensure its suitability for integration into a knowledge graph. This includes removing duplicates and converting the data into a format suitable for network analysis.
3. Knowledge graph completion: With the preprocessed data, the tutorial moves into applying the TransE model, a popular method in knowledge graph completion. TransE models relations as translations on a low-dimensional embedding of the entities, making it easier to infer missing links between drugs and potential new disease targets.
4. Evaluation: The final phase involves evaluating the predictions made by the knowledge graph. This is done using a time-slicing technique where the model's predictions are compared against known drug-disease relationships that emerge after the data cutoff date. This method helps validate the effectiveness of the model and provides a measure of how predictive and useful the model could be in real-world scenarios.

References

Altay, O., Mohammadi, E., Lam, S., Turkez, H., Boren, J., Nielsen, J., Uhlen, M., & Mardinoglu, A. (2020). Current status of COVID-19 therapies and drug repositioning applications. *Iscience*, 101303.

Boccaletti, S., Latora, V., Moreno, Y., Chavez, M., & Hwang, D. (2006). Complex networks: Structure and dynamics. *Physics Reports*, *424*(4), 175–308.

Bojanowski, P., Grave, E., Joulin, A., & Mikolov, T. (2017). Enriching word vectors with subword information. *Transactions of the Association for Computational Linguistics*, *5*, 135–146.

Bordes, A., Usunier, N., Garcia-Duran, A., Weston, J., & Yakhnenko, O. (2013). Translating embeddings for modeling multi-relational data. In C. J. C. Burges, L. Bottou, M. Welling, Z. Ghahramani, & K. Q. Weinberger (Eds.), *Advances in neural information processing systems 26* (pp. 2787–2795). Curran Associates, Inc.

Cestnik, B., Fabbretti, E., Gubiani, D., Urbančič, T., & Lavrač, N. (2017). Reducing the search space in literature-based discovery by exploring outlier documents: A case study in finding links between gut microbiome and Alzheimer's disease. *Genomics and Computational Biology*, *3*(3), e58.

Conneau, A., Lample, G., Ranzato, M., Denoyer, L., & Jégou, H. (2017). Word translation without parallel data. *Computing Research Repository (CoRR)*. http://arxiv.org/abs/1710.04087

Ge, Y., Tian, T., Huang, S., Wan, F., Li, J., Li, S., Yang, H., Hong, L., Wu, N., Yuan, E., Cheng, L., Lei, Y., Shu, H., Feng, X., Jiang, Z., Chi, Y., Guo, X., Cui, L., Xiao, L., … Zeng, J. (2020). A data-driven drug repositioning framework discovered a potential therapeutic agent targeting COVID-19. *bioRxiv*.

Hinton, G. E., McClelland, J. L., & Rumelhart, D. E. (1986). Distributed representations. In D. E. Rumelhart, J. L. McClelland, & PDP Research Group (Eds.), *Parallel distributed processing, volume 1: Explorations in the microstructure of cognition: Foundations* (pp. 77–109, Vol. 1). The MIT PRess.

Juršič, M., Cestnik, B., Urbančič, T., & Lavrač, N. (2012). Cross-domain literature mining: Finding bridging concepts with CrossBee. *Proceedings of the 3rd International Conference on Computational Creativity*, 33–40.

Kilicoglu, H., Shin, D., Fiszman, M., Rosemblat, G., & Rindflesch, T. C. (2012). SemMedDB: A PubMed-scale repository of biomedical semantic predications. *Bioinformatics*, *28*(23), 3158–3160.

Lavrač, N., Martinc, M., Pollak, S., Pompe-Novak, M., & Cestnik, B. (2020). Bisociative literature-based discovery: Lessons learned and new word embedding approach. *New Generation Computing*, *38*(4), 773–800.

McInnes, B. T. (2004, December). *Extending the log-likelihood measure to improve collocation identification* [Master's thesis, Univerity of Minnesota].

Mikolov, T., Chen, K., Corrado, G., & Dean, J. (2013). Efficient estimation of word representations in vector space. *arXiv preprint 1301.3781*.

Mikolov, T., Le, Q. V., & Sutskever, I. (2013). Exploiting similarities among languages for machine translation. *arXiv preprint 1309.4168*.

Pennington, J., Socher, R., & Manning, C. (2014). GloVe: Global vectors for word representation. *Proceedings of Empirical Methods in Natural Language Processing, EMNLP*, 1532–1543.

Petrič, I., Cestnik, B., Lavrač, N., & Urbančič, T. (2012). Outlier detection in cross-context link discovery for creative literature mining. *The Computer Journal*, *55*(1), 47–61.

Pushpakom, S., Iorio, F., Eyers, P. A., Escott, K. J., Hopper, S., Wells, A., Doig, A., Guilliams, T., Latimer, J., McNamee, C., et al. (2019). Drug repurposing: Progress, challenges and recommendations. *Nature Reviews Drug Discovery*, *18*(1), 41–58.

Sluban, B., Juršič, M., Cestnik, B., & Lavrač, N. (2012). Exploring the power of outliers for cross-domain literature mining. In M. R. Berthold (Ed.), *Bisociative Knowledge Discovery* (pp. 325–337). Springer.

Sun, Z., Deng, Z., Nie, J., & Tang, J. (2019). RotatE: Knowledge graph embedding by relational rotation in complex space. *arXiv*. http://arxiv.org/abs/1902.10197

Trouillon, T., Welbl, J., Riedel, S., Gaussier, É., & Bouchard, G. (2016). Complex embeddings for simple link prediction.

Wang, X., & Guan, Y. (2020). COVID-19 drug repurposing: A review of computational screening methods, clinical trials, and protein interaction assays. *Medicinal Research Reviews*.

Yang, B., Yih, W.-t., He, X., Gao, J., & Deng, L. (2014). Embedding entities and relations for learning and inference in knowledge bases. *arXiv preprint arXiv:1412.6575*.

Zeng, X., Song, X., Ma, T., Pan, X., Zhou, Y., Hou, Y., Zhang, Z., Li, K., Karypis, G., & Cheng, F. (2020). Repurpose open data to discover therapeutics for covid-19 using deep learning. *Journal of proteome research*.

Zhang, R., Hristovski, D., Schutte, D., Kastrin, A., Fiszman, M., & Kilicoglu, H. (2021). Drug repurposing for covid-19 via knowledge graph completion. *Journal of biomedical informatics*, *115*, 103696.

Zhou, Y., Hou, Y., Shen, J., Huang, Y., Martin, W., & Cheng, F. (2020). Network-based drug repurposing for novel coronavirus 2019-ncov/sars-cov-2. *Cell Discovery*, *6*(1), 1–18.

Zhou, Y., Wang, F., Tang, J., Nussinov, R., & Cheng, F. (2020). Artificial intelligence in COVID-19 drug repurposing. *The Lancet Digital Health*.

Chapter 10
Research Trends and Lessons Learned

This chapter starts by outlining some lessons learned in our previous LBD research in Section 10.1.[1] Next, it presents new research avenues opened by the advent of Large Language Models (LLMs) in Section 10.2, followed by a mini-study on the use of LLMs for LBD domain conceptualization in Section 10.3. The chapter concludes with Section 10.4, which includes a Python tutorial allowing for simplified LBD experiment replicability and code reuse.

10.1 Selected lessons learned from past LBD research

Much of our past work was based on exploiting the information hidden in outlier documents to reduce the search for b-terms. Outliers represent unexpected events, entities, items, or documents. Our early research in LBD has focused on the identification and exploration of outlier documents, since they frequently embody new information that is often hard to explain in the context of existing mainstream knowledge. Our LBD research described in the papers by Petrič et al. (2012) and Sluban et al. (2012) suggested that bridging terms are more frequent in documents that are in some sense different from the majority of documents in a given domain.

The outlier-based approach to LBD proposed by Petrič et al. (2012) used document clustering to find outlier documents. The approach consists of two steps. In the first step, the OntoGen clustering algorithm (Fortuna et al., 2006) was applied to cluster a merged document set A ∪ C, consisting of documents from two domains A and C. Unsupervised clustering results in two document clusters: A′ = *Classified as A* (i.e. documents from A ∪ C classified as A), and C′= *Classified as C* (i.e. documents from A ∪ C classified as C). In the second step of outlier detection, the groups A′ and C′ were further separated, each into two clusters, based on the original labels of the documents A and C. As a result, a two-level tree hierarchy of clusters was generated, illustrated in Figure 6.4. For convenience, it is shown again in Figure 10.1.

[1] This section is based on the work published by Lavrač et al., 2020.

N. Lavrač et al., *Bisociative Literature-Based Discovery*,
https://doi.org/10.1007/978-3-031-96863-1_10

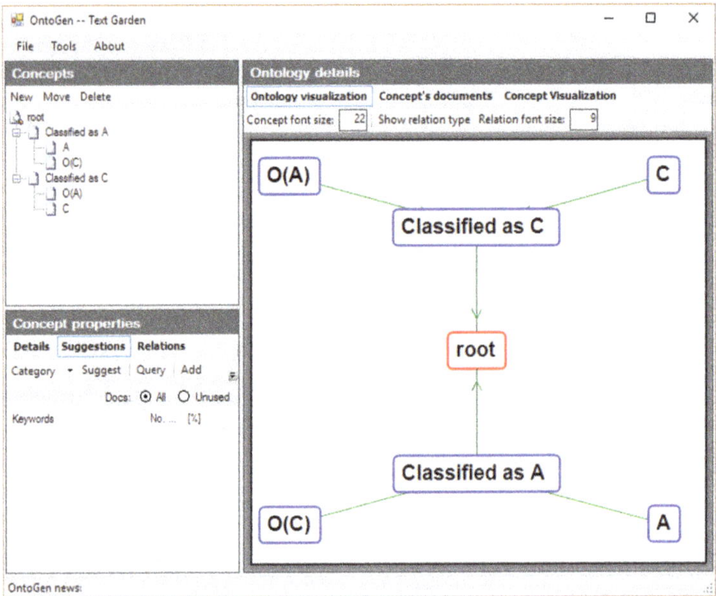

Fig. 10.1 Target domain documents from literatures A and C, clustered according to OntoGen's two-step approach, first using unsupervised and then supervised clustering to obtain outlier documents O(A) and O(C) of literatures A and C, respectively.

First Lesson Learned. The hypothesis that outlier documents have the potential to improve the effectiveness of bridging term detection was tested on data sets from domain pairs *migraine-magnesium* (Swanson et al., 2006) and *autism-calcineurin* (Petrič et al., 2009), which have lists of confirmed concept bridging terms (b-terms). The experimental results obtained by using OntoGen confirm the hypothesis that most bridging terms appear in outlier documents and that by considering only outlier documents the search space for b-term identification can be largely reduced.

In our past work (Sluban et al., 2012), we proposed to detect outlier documents using classification algorithms for classification noise filtering, first suggested by Brodley and Friedl (1999). Having documents from two domains of interest A and C, Sluban et al. (2012) first trained an ensemble classifier that distinguishes between the documents of these domains and uses the ensemble classifier to classify all the documents. Misclassified documents are declared as outliers, since according to the classification model, they do not belong to their domain (class label) of origin. These outliers can be interpreted as borderline documents as they were considered by the model to be more similar to the other domain than to their original domain, and can be regarded as bridging documents between the two domains. In other words, if an instance of class A is classified in the opposite class C, we consider it an outlier of domain A, and vice versa, where the two sets of outlier documents denoted with O(A) and O(C) were illustrated in Figure 6.1 in Section 6.2.

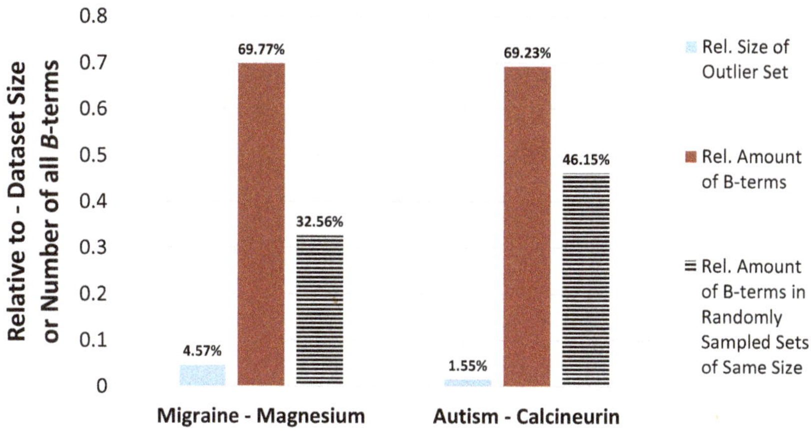

Fig. 10.2 Presence of b-terms in the detected outlier sets of two domain pair datasets.

Second Lesson Learned. The hypothesis that outlier documents have the potential to improve the effectiveness of bridging term detection using a classification filtering approach to outlier detection by Sluban et al. (2012) was again tested on datasets from domain pairs *migraine-magnesium* (Swanson et al., 2006) and *autism-calcineurin* (Petrič et al., 2009), which have lists of confirmed bridging terms (b terms). The experimental results showed that the sets of detected outlier documents were relatively small (including less than 5% of entire datasets) and that they contained a large majority of bridging terms, which was significantly higher than in the same-sized random subsets. Hence, the effort needed for finding cross-domain links is substantially reduced, as it requires exploring a much smaller subset of documents, where a great majority of b-terms are present and more frequent. These results were summarized in Figure 6.2 in Section 6.2. For convenience, it is shown again in Figure 10.2.

When applied to a new domain, in this case the *Alzheimer's disease-gut microbiome* domain pair (Cestnik et al., 2017), the OntoGen method uses domains A and C, and builds a joint set of documents $A \cup C$. With this intention, two individual sets of documents (e.g., titles, abstracts or full texts of scientific articles), one for each researched domain (namely, literature A on Alzheimer's disease and literature C on the gut microbiome), were automatically retrieved from the PubMed database. A cluster hierarchy was constructed from the dataset of 17,863 papers with OntoGen. Two first-level clusters are labeled using OntoGen suggested keywords *ad, abeta, cognitive* and *microbiota, gut, intestine*. Four second-level sub-clusters separate documents according to their original search keywords for Alzheimer's disease and the gut microbiome, as illustrated in Figure 10.3.

Third Lesson Learned. In Alzheimer's disease-gut microbiome LBD application, the initial document set $A \cup C$ consisted of some documents at the intersection of A and C, which means that some documents were retrieved from PubMed by separate queries for the domains A (*Alzheimer*) and C ((*gut* OR *intestinal*) AND

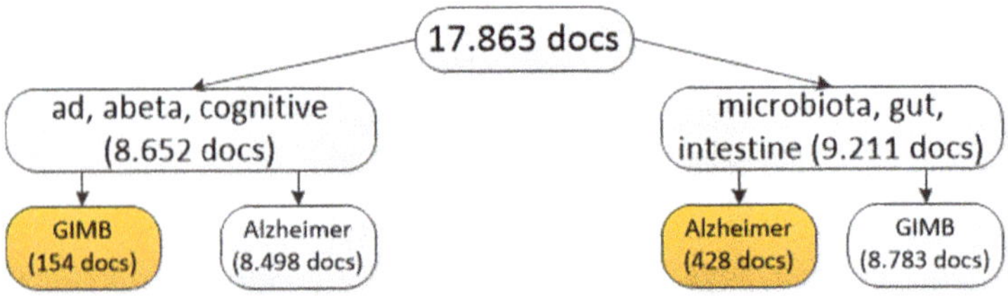

Fig. 10.3 Two-level cluster hierarchy constructed with OntoGen from the dataset of 17,863 papers in the *Alzheimer's disease-gut microbiome* domain pair.

(*microbiota* OR *bacteria*)), which was surprising. After carefully inspecting these documents (as these documents could contain the b-terms representing a solution to the problem, which proved not to be the case), we realized that keeping them in the $A \cup C$ document set was problematic, causing the methodology to fail due to biased calculations of heuristics. As a result, we eliminated the documents that were retrieved by both queries, resulting in 17,863 documents kept in the set of documents $A \cup C$ used for further exploration.

Fourth Lesson Learned. In Alzheimer's disease-gut microbiome LBD application using OntoGen for the detection of outlier documents, the space of documents used for the b-term exploration was further reduced from the set of 17,863 documents to only 154 papers on gut microbiome and 428 papers on Alzheimer's disease. In the b-term exploration, we considered all terms and not just medical ones. A list of 523 English stop words was used to filter out meaningless words, and English Porter stemming was applied, which helped us to find medically interesting b-terms.

Fifth Lesson Learned. By processing the remaining 582 outlier documents, we used CrossBee (Juršič et al., 2012) to extract 4,723 terms as potential b-terms connecting the two domains. Although the list of potential bridging terms was ordered according to the bridging term potential estimated by the ensemble heuristic, browsing and analyzing the terms from the list still presented a substantial burden to the domain expert and the development team. To further reduce the size of the potential b-term list, the domain expert prepared his own list of 289 interesting terms. It included common terms and specific molecular factors and pathways, which were manually identified in titles, abstracts, and keywords of 42 articles obtained from the PubMed search query (*gut AND Alzheimer*), 55 of which also appeared among the 4,723 terms extracted by CrossBee. During the evaluation phase, the relevant papers for each b-term candidate were reviewed and searched for potential clues for further investigation.

Compared to outlier document detection using OntoGen, an upgraded methodology proposed by Cestnik et al. (2017) was implemented in a reusable outlier-based LBD methodology within a web-based text mining platform TextFlows[2] (Perovšek et al., 2016) that allowed us to construct and execute advanced text mining workflows. The workflow shown in Figure 6.6 (repeated for convenience in Figure 10.4) consists of seven

[2] http://textflows.org

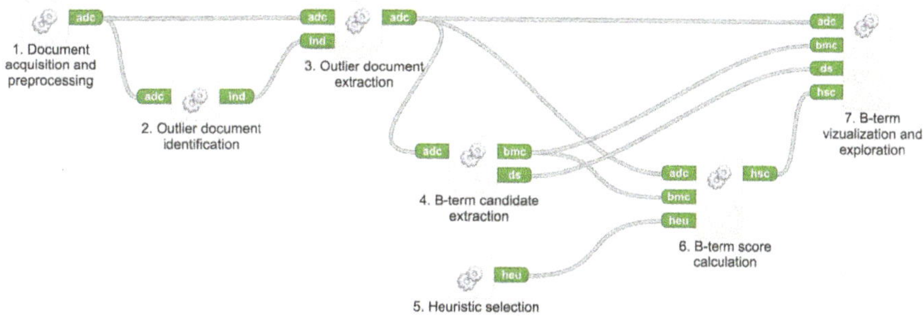

Fig. 10.4 A top-level workflow of the LBD methodology in TextFlows (Perovšek et al., 2016).

steps implemented as subprocesses. The connections between subprocesses represent the flow of documents from one subprocess to another. In summary, steps 1-3 represent outlier detection and steps 4-7 represent cross-domain exploration for b-term detection.

Sixth Lesson Learned. In the experiments in the TextFlows environment, we used the NoiseRank ensemble-based outlier detection approach developed by Sluban et al. (2014) as implemented in TextFlows. The goal of the first three steps of the methodology is to effectively extract a set of outlier documents from the whole corpus of input documents. Consequently, by decreasing the size of the input set of documents, the second phase becomes more focused, efficient, and effective. In the last four steps of the workflow in Figure 6.6, the components of the CrossBee HCI interface (**jurvsivc2012cross**) are executed to support expert-guided b-term analysis. Here, the goal is to further prepare the input documents for b-term visualization and exploration. Note that in this step, the role of the domain expert is crucial.

10.2 Towards LBD using Large Language Models

An important family of Large Language Models (LLMs) is Masked Language Models (MLMs) such as BERT (Devlin et al., 2019). This research was followed by the development of sentence transformers such as SentenceBERT, which are adapted to the task of learning sentence representations (Reimers & Gurevych, 2019)), topic modeling approaches such as BERTopic (Grootendorst, 2022), and keyword extraction approaches such as KeyBERT (Grootendorst, 2020)). Another family consists of generative autoregressive language models such as LLaMa2 (Touvron et al., 2023). Language models have established new standards for understanding and generating texts with human precision (Min et al., 2023).

Language models can profit from in-context learning, crucial for eliciting accurate and specific output from language models, such as LLaMa2 (Touvron et al., 2023), demonstrating an advanced level of understanding and flexibility in language generation

(Pan et al., 2023). The in-context learning approach involves crafting specific prompts that enable models to learn from, e.g., a single example within the prompt context, thereby generating more relevant and contextually nuanced responses. LLM prompting techniques (Wei et al., 2022), used for in-context learning of LLMs can perform new tasks by adapting to one (or a few) in-context example(s) (Lampinen et al., 2022).

Recent work of Q. Wang et al. (2023) has shown the benefit of using LLMs in Literature-Based Discovery (LBD). The authors argue that standard LBD is limited to predicting pairwise relations between discrete concepts, such as drug-disease links, and that LBD also ignores critical contexts, such as experimental settings (e.g., a specific patient population where a drug is evaluated) and background knowledge and motivations that human scientists consider (e.g., to find a drug candidate without specific side effects). The authors address these limitations with a novel formulation of contextualized-LBD (C-LBD), i.e. generating scientific hypotheses in a natural language while grounding them in a context that controls the hypothesis search space. They present a new modeling framework using retrieval of "inspirations" from a heterogeneous network of citations and knowledge graph relations and create a new dataset, derived from papers. In automated and human evaluations, their models improve over powerful LLM baselines, but also reveal challenges on the road to building machines that generate new scientific knowledge.

Koneru et al. (2024) explored the ability of LLMs to recognize evidence supporting or refuting specific hypotheses based on the text of scientific abstracts. They present a public dataset for the task of proving scientific hypotheses; however, this dataset is not associated with the LBD domain. Ciucă et al. (2023) investigated the power of adversarial prompting and LLMs for robust hypothesis generation, applying it to the field of astronomy. Yang et al. (2024) investigate LLMs for automated open-domain scientific hypotheses discovery, with the goal of developing a system that automatically generates valid, novel, and helpful hypotheses, given only a set of raw web corpora. The authors have developed a multimodule framework with feedback mechanisms that shows performance improvements compared to a GPT-4 based evaluation and a social science expert evaluation. Finally, Lahav et al. (2022) presented a task to extract and search scientific challenges and directions to facilitate knowledge discovery, focusing on interdisciplinary work related to the COVID-19 pandemic. The authors created an expert-annotated corpus of texts labeled with semantic categories, trained a model to identify new challenges and directions, and developed a specialized search engine that outperforms a popular scientific search engine in helping to discover new knowledge.

10.3 Using LLMs for LBD literature structuring

This section presents the AHAM (Adapt, Help, Ask, Model) methodology and score, developed by Koloski et al. (2024). The AHAM methodology uses the LLaMa2 model to generate topic definitions through one-shot learning, with the **help** from domain experts to craft prompts that guide literature mining by **asking** the model to label topics discovered by the BERTopic topic **model**ing framework. In summary, the proposed methodology consists of the following steps:

Adapt: Domain adaptation (see Section 10.3.1).
Help : Crafting a prompt with the help of domain experts for topic modeling (see Section 10.3.2).
Ask : Prompting the model with the hand-crafted prompt (see Section 10.3.2).
Model : Modeling the topics by BERTopic (see Section 10.3.3).

An integral part of the methodology is the novel AHAM heuristic (presented in Section 10.3.4), which employs language generation and translation scores for inter-topic similarity assessment to minimize outlier topics and overlap between topic definitions.

The AHAM methodology was experimentally assessed on the LBD corpus of 389 publications spanning from 1986 to 2023, which was briefly outlined in Section 2.3.4, where we quantitatively evaluated the AHAM heuristic, and qualitatively assessed the quality of discovered topics and topic names.

10.3.1 Domain-adaptation via sentence-transformers

Contrastive learning is a deep learning technique for unsupervised representation learning. Its goal is to learn a representation of data such that similar instances are close together in the representation space, while dissimilar instances are far apart. Sentence transformer models (Reimers & Gurevych, 2019) are pre-trained using a contrastive learning objective function. For a query q, a positive sentence p^+ (similar to q), and a negative sentence p^- (different from q), the loss function L is defined as:

$$L(q, p^+, p^-) = \max(0, f(q, p^+) - f(q, p^-) + margin),$$

where f measures the similarity of sentences (often with the cosine similarity function), and *margin* is a hyperparameter for the minimum positive-negative pair distance.

Sentence transformers are initially pre-trained on general domains D_s[3] and may lack specific tuning for a particular domain of interest D_t. To overcome this limitation, K. Wang et al. (2022) proposed a multistep adaptation process, named *Domain Adaptation via Generative Pseudo Labeling (GPL)*. In this process, a variant of the T5 Text-to-Text Transfer Transformer (Raffel et al., 2020) is used to generate synthetic queries q_i, forming $Q_{\text{synth}} = \text{T5}(D_s)$.

Mining Negative Passages then finds various negative sentences P_{neg} for each query q_i using pre-trained models and nucleus sampling (which, from a given set of words, chooses the smallest possible set of words whose cumulative probability exceeds the given probability). Training triplets $t = (q_i, p_i^+, p_i^-)$ are constructed and a cross encoder CE scores each pair (q_i, p_i^+) and (q_i, p_i^-) as $\delta_{t_i} = CE(q_i, p_i^+) - CE(q_i, p_i^-)$, creating a training corpus to tune a given model S to D_t with the MarginMSE loss (Hofstätter et al., 2020).

[3] In this study, pre-training was performed by the S = *all-mini-LM-v12* sentence transformer model (Reimers & Gurevych, 2019).

10.3.2 Prompt engineering of LLMs to design topic names

We utilize the LLaMa2 language model, specifically the *Llama-2-13b-chat-hf* variant from the HuggingFace[4] repository, setting the temperature to 0, the context window to 500 tokens, and the repetition penalty to 1.1. The objective of this phase is to semantically utilize the extracted keywords and the most central documents in each topical cluster. We aim to leverage the LLM's capabilities to derive meaningful semantic labels. This is achieved through training via one-example in-context learning, guided by prompt designs crafted by domain experts of the meta-literature analysis application of our interest. Following related work, we have engineered a three-level prompt structure:

- System prompt, to give personality to the LLM for labeling topics as a domain expert:

  > You are a helpful, respectful, and honest research assistant for labeling topics.

- One shot, example prompt, from which we **help** the model with guidance for in-context learning. Notice that this prompt has been crafted in a specific manner, reflecting our contributions to the field, but it could have been crafted differently by the end user.

  > I have a topic that contains the following documents:
  > - Bisociative Knowledge Discovery by Literature Outlier Detection.
  > - Evaluating Outliers for Cross-Context Link Discovery.
  > - Exploring the Power of Outliers for Cross-Domain Literature Mining.
  > The topic is described by the following keywords: bisociative, knowledge discovery, outlier detection, data mining, cross-context, link discovery, cross-domain, machine learning.
  > Based on the information about the topic above, please create a simple, short, and concise computer science label for this topic. Make sure you only return the label and nothing more.
  > [INST]: Outlier-based knowledge discovery

- The query prompt, which **asks** the model to label topics from the keywords and the most central documents for a given topic, is personalized. This personalization ensures that the model returns computer-science topics from the field of Literature-Based Discovery without explicitly mentioning the topics:

  > I have a topic that contains the following documents [DOCUMENTS]
  > The topic is described by the following keywords: [KEYWORDS]
  > Based on the information about the topic above, please create a simple, short and concise computer science label for this topic. Make sure you only return the label and nothing more.

10.3.3 BERTopic modeling

The initial step in BERTopic is to reduce the dimensionality of the adapted sentence-transformer embeddings using UMAP (McInnes et al., 2018) to facilitate more efficient

[4] https://huggingface.co/

clustering, as clustering in high-dimensional spaces can be challenging. The UMAP configuration involves setting the number of the nearest neighbors and components (set to 5), with a minimum distance (set to 0.0), and adopting cosine similarity as the score. UMSP dimensionality reduction is followed by the application of the HDBSCAN clustering algorithm (McInnes et al., 2017) to identify distinct clusters. For HDBSCAN, the minimum cluster size was set to 5, with all other parameters left at their default values.

The method then progresses to extracting document-level features to characterize the vicinity of each topic, initially employing class-based cTFIDF to assess word significance at the corpus level, thereby accentuating their importance to specific topics. Subsequently, for each cluster, the KeyBERT keyword extraction technique, which uses the S sentence-transformer representation, is applied to retrieve and rank relevant keywords.

The final output, obtained at the end of the processing pipeline for a set of N documents, is a set of k distinct clusters. Among these, a unique cluster named k_{outliers} is defined, representing outlier documents that did not closely correspond to any other cluster.

10.3.4 The AHAM heuristic

To assess the effect of domain adaptation on topic modeling, we used the idea that, on average, the names of topic labels tend to be dissimilar. To measure the numerical similarities between topics, we have chosen to use three distinct *TopicSimilarity* scores to evaluate the similarity of topics named Topic A and Topic B:

Levenstein fuzzy matching: We employ the *fuzzywuzzy*[5] Python implementation that utilizes normalization based on length to provide a normalized *lexical* similarity score between the topic names.

BERTscore: Uses BERT model embeddings to evaluate the similarity between A and B, by comparing the semantic embedding similarity of the present n-grams w. BERTscore is defined as: $\text{BERTscore} = \frac{1}{|A|} \cdot \sum_{w \in \text{A}} \max_{w' \in \text{B}} \cos(w, w')$.

Semantic similarity (using all-mini-LM-v12): Evaluates the semantic closeness of A and B using the cosine of the angle between their vector representations, S(A) and S(B), defined as: $\cos(\text{A}, \text{B}) = \frac{S(A) \cdot S(B)}{\|S(A)\| \cdot \|S(B)\|}$.

To assess adaptation through the evaluation of topic naming, we propose a heuristic that jointly combines the number of outliers k_{outliers}, the number of clusters k, and the similarity between the generated topic names with the *TopicSimilarity* score to the number of steps n_{steps} in domain adaptation. The $\texttt{AHAM}_{\texttt{objective}}$ is defined as follows:

$$\texttt{AHAM}_{\texttt{objective}} = 2 \cdot \frac{k_{\text{outliers}}}{k} \cdot \sum_{\substack{i=1 \\ j=i+1}}^{k} \frac{\texttt{TopicSimilarity}(k_i, k_j)}{k(k-1)}.$$

We evaluated on every 10,000 step of the GPL and selected the topic modeling that minimizes the $\texttt{AHAM}_{\texttt{objective}}$ score within some evaluation budget of n_{steps}.

[5] https://github.com/seatgeek/fuzzywuzzy

10.3.5 Qualitative evaluation and future directions

We validated the AHAM methodology on the LBD corpus of 389 publications spanning from 1986 to 2023, which was briefly outlined in Section 2.3.4. We observed that document partitioning via domain adaptation significantly improves cluster validity. Manual evaluation by literature-based discovery (LBD) experts of clustering outcomes (i.e., the baseline, arXiv-specific, and medrXiv-specific models) revealed distinct patterns. The baseline model identified a broad "Discovery through Statistical and Knowledge-Based Methods" cluster, while domain-adapted models resulted in more focused clusters like "BioMed Text Mining" and "Knowledge Linkage Discovery," alongside specialized topics such as "Kostoff's cluster" and "Biomedical Connections Discovery." Figure 10.5 presents topic evolution between the initial topics and the topics identified after the domain adaptation selected by AHAM.

Future directions include extensive testing of domain adaptation's efficacy in topic modeling across various fields, leveraging the AHAM objective. Further, we aim to assess the benefits of in-domain data training, explore model adaptability across diverse languages, particularly those with limited resources, and investigate the influence of the size of the generative language model and different sentence transformers on topic identification.

10.4 Tutorial in Python

The content presented in this chapter is supplemented by a practical example, the source code for which is hosted on GitHub. We recommend that readers access the tutorial **09_aham_topic_modeling.ipynb** in the book's GitHub repository to further explore the concepts discussed herein. Before running the Jupyter notebooks, readers should refer to Chapter 4 and follow the instructions on how to install Python and the required packages from the **requirements.txt** file in the repository.

10.4.1 09_aham_topic_modeling

This tutorial demonstrates topic modeling on a corpus of LBD literature using the Adapt, Help, Ask, Model (AHAM) methodology (Koloski et al., 2024). The AHAM methodology guides the domain-specific adaptation of the BERTopic (Grootendorst, 2022) topic modeling framework to improve scientific text analysis. By utilizing the LLaMa2 generative language model, we generate topic definitions via one-shot learning by crafting prompts with the help of domain experts to guide the LLM for literature mining by asking it to model the topic names.

The workflow utilizes tools and methods from the AHAM framework, incorporating advanced natural language processing techniques and domain-specific adaptations to enhance topic discovery.

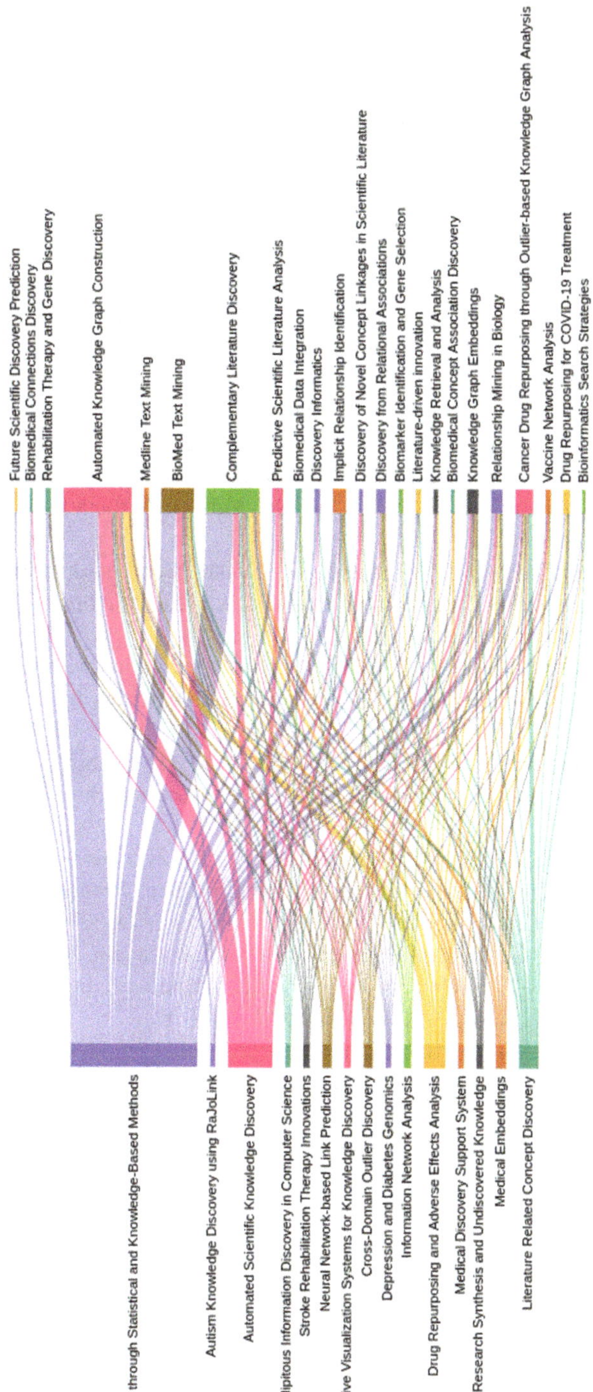

Fig. 10.5 Evolution of topics between the first and best step (after 40 steps) of domain adaptation following AHAM objective, on the arXiv domain.

References

Brodley, C. E., & Friedl, M. A. (1999). Identifying mislabeled training data. *Journal of Artificial Intelligence Research, 11*, 131–167.

Cestnik, B., Fabbretti, E., Gubiani, D., Urbančič, T., & Lavrač, N. (2017). Reducing the search space in literature-based discovery by exploring outlier documents: A case study in finding links between gut microbiome and Alzheimer's disease. *Genomics and Computational Biology, 3*(3), e58.

Ciucă, I., Ting, Y.-S., Kruk, S., & Iyer, K. (2023). Harnessing the power of adversarial prompting and large language models for robust hypothesis generation in astronomy. https://arxiv.org/abs/2306.11648

Devlin, J., Chang, M.-W., Lee, K., & Toutanova, K. (2019, June). BERT: Pre-training of deep bidirectional transformers for language understanding. In J. Burstein, C. Doran, & T. Solorio (Eds.), *Proceedings of the 2019 Conference of the North American Chapter of the Association for Computational Linguistics: Human Language Technologies* (pp. 4171–4186, Vol. 1). Association for Computational Linguistics.

Fortuna, B., Grobelnik, M., & Mladenić, D. (2006). Semi-automatic data-driven ontology construction system. *Proceedings of the 9th International Multiconference Information Society*, 223–226.

Grootendorst, M. (2020). KeyBERT: Minimal keyword extraction with BERT.

Grootendorst, M. (2022). BERTopic: Neural topic modeling with a class-based TF-IDF procedure.

Hofstätter, S., Althammer, S., Schröder, M., Sertkan, M., & Hanbury, A. (2020). Improving efficient neural ranking models with cross-architecture knowledge distillation.

Juršič, M., Cestnik, B., Urbančič, T., & Lavrač, N. (2012). Cross-domain literature mining: Finding bridging concepts with CrossBee. *Proceedings of the 3rd International Conference on Computational Creativity*, 33–40.

Koloski, B., Lavrač, N., Cestnik, B., Pollak, S., Škrlj, B., & Kastrin, A. (2024). AHAM: Adapt, Help, Ask, Model Harvesting LLMs for literature mining. *Advances in Intelligent Data Analysis XXII. IDA 2024.*, 254–265.

Koneru, S., Wu, J., & Rajtmajer, S. (2024). Can large language models discern evidence for scientific hypotheses? case studies in the social sciences. https://arxiv.org/abs/2309.06578

Lahav, D., Falcon, J. S., Kuehl, B., Johnson, S., Parasa, S., Shomron, N., Chau, D. H., Yang, D., Horvitz, E., Weld, D. S., & Hope, T. (2022). A search engine for discovery of scientific challenges and directions. https://arxiv.org/abs/2108.13751

Lampinen, A., Dasgupta, I., Chan, S., Mathewson, K., Tessler, M., Creswell, A., McClelland, J., Wang, J., & Hill, F. (2022). Can language models learn from explanations in context? In Y. Goldberg, Z. Kozareva, & Y. Zhang (Eds.), *Findings of the Association for Computational Linguistics: EMNLP 2022* (pp. 537–563). Association for Computational Linguistics.

Lavrač, N., Martinc, M., Pollak, S., Pompe-Novak, M., & Cestnik, B. (2020). Bisociative literature-based discovery: Lessons learned and new word embedding approach. *New Generation Computing*, *38*(4), 773–800.

McInnes, L., Healy, J., & Astels, S. (2017). HDBSCAN: Hierarchical density based clustering. *Journal of Open Source Software*, *2*(11), 205.

McInnes, L., Healy, J., Saul, N., & Großberger, L. (2018). UMAP: Uniform manifold approximation and projection. *Journal of Open Source Software*, *3*(29), 861.

Min, B., Ross, H., Sulem, E., Veyseh, A. P. B., Nguyen, T. H., Sainz, O., Agirre, E., Heintz, I., & Roth, D. (2023). Recent advances in natural language processing via large pre-trained language models: A survey. *ACM Computing Surveys*, *56*(2), 1–40.

Pan, J., Gao, T., Chen, H., & Chen, D. (2023). What in-context learning "learns" in-context: Disentangling task recognition and task learning. In A. Rogers, J. Boyd-Graber, & N. Okazaki (Eds.), *Findings of the Association for Computational Linguistics: ACL 2023* (pp. 8298–8319). Association for Computational Linguistics.

Perovšek, M., Kranjc, J., Erjavec, T., Cestnik, B., & Lavrač, N. (2016). Textflows: A visual programming platform for text mining and natural language processing. *Special Issue on Knowledge-based Software Engineering, Science of Computer Programming*, *121*, 128–152.

Petrič, I., Cestnik, B., Lavrač, N., & Urbančič, T. (2012). Outlier detection in cross-context link discovery for creative literature mining. *The Computer Journal*, *55*(1), 47–61.

Petrič, I., Urbančič, T., Cestnik, B., & Macedoni-Lukšič, M. (2009). Literature mining method RaJoLink for uncovering relations between biomedical concepts. *Journal of Biomedical Informatics*, *42*(2), 219–227.

Raffel, C., Shazeer, N., Roberts, A., Lee, K., Narang, S., Matena, M., Zhou, Y., Li, W., & Liu, P. J. (2020). Exploring the limits of transfer learning with a unified text-to-text transformer. *Journal of Machine Learning Research*, *21*(140), 1–67.

Reimers, N., & Gurevych, I. (2019). Sentence-BERT: Sentence embeddings using siamese BERT-networks. *Proceedings of the 2019 Conference on Empirical Methods in Natural Language Processing*. https://arxiv.org/abs/1908.10084

Sluban, B., Juršič, M., Cestnik, B., & Lavrač, N. (2012). Exploring the power of outliers for cross-domain literature mining. In M. R. Berthold (Ed.), *Bisociative Knowledge Discovery* (pp. 325–337). Springer.

Sluban, B., Gamberger, D., & Lavrač, N. (2014). Ensemble-based noise detection: Noise ranking and visual performance evaluation. *Data Mining and Knowledge Discovery*, *28*(2), 265–303.

Swanson, D. R., Smalheiser, N. R., & Torvik, V. I. (2006). Ranking indirect connections in literature-based discovery: The role of Medical Subject Headings (MeSH). *Journal of the American Society for Information Science and Technology*, *57*(11), 1427–1439.

Touvron, H., Martin, L., Stone, K., Albert, P., Almahairi, A., & Babaei, Y. (2023). *Llama 2: Open Foundation and Fine-Tuned Chat Models*. ArXiv. https://doi.org/10.48550/arXiv.2307.09288

Wang, K., Thakur, N., Reimers, N., & Gurevych, I. (2022). GPL: Generative pseudo labeling for unsupervised domain adaptation of dense retrieval. *Proceedings of the 2022 Conference of the North American Chapter of the Association for Computational Linguistics: Human Language Technologies*, 2345–2360.

Wang, Q., Downey, D., Ji, H., & Hope, T. (2023). Learning to generate novel scientific directions with contextualized literature-based discovery.

Wei, J., Wang, X., Schuurmans, D., Bosma, M., Xia, F., Chi, E., Le, Q. V., Zhou, D., et al. (2022). Chain-of-thought prompting elicits reasoning in large language models. *Advances in Neural Information Processing Systems*, *35*, 24824–24837.

Yang, Z., Du, X., Li, J., Zheng, J., Poria, S., & Cambria, E. (2024). Large language models for automated open-domain scientific hypotheses discovery. https://arxiv.org/abs/2309.02726

Index

ABC model, 2, 4
accuracy, 57
Alzheimer's disease, 104
AnC model, 23
Area Under ROC, 84
Arrowsmith, 4, 80
artificial neural network, 53
association, 10, 129
  cross-context, 2
AUC, 84
autism, 121

bag of words, 47, 48
benchmark data, 67
BERT, 54
bisociation, 2, 10
bisociation score, 82
BoW, *see* bag of words
bridge, 3
bridging link, 3
bridging term, 3

calcineurin, 121
circadian rhythm, 145
classification error, 57
closed discovery, 4, 77, 129, 143
clustering, 96
CNN, *see* convolutional neural network
co-occurrence, 129
community detection, 52
complex networks, 129
concept, 2
concept representation, 2
convolutional neural network, 54
corpus, 46
cosine similarity, 49
creativity support tool, 7, 61
cross context association, 2

CrossBee, 82, 161
CrossBee
  heuristics, 83
  interface, 88
  methodology, 82
CST, *see* creativity support tool

deep neural network, 53
dependency parsing, 47
discovery
  closed, 4
  literature-based, 1
  open, 5
distance measure, 49, 50
DistMult, 152
DNN, *see* deep neural network
document, 46
domain adaptation, 55
drug repurposing, 149

ELMo, 54
embedding, 48, 54, 141
embeddings
  learning, 141
ensemble heuristics, 86
error
  classification, 57
evaluation, 57
evaluation
  accuracy, 57

F-measure, 58
false negative, 57
false positive, 57
false positive rate, 58
FastText, 142
feature, 46
fish oil, 111, 115

N. Lavrač et al., *Bisociative Literature-Based Discovery*,
https://doi.org/10.1007/978-3-031-96863-1

Gene Ontology, 35, 51
GloVe, 142
GO, see Gene Ontology35, 51
gold standard, 68
graph, 130
gut microbiota, 104

HCI, see human-computer interaction7
heuristics, 83
k, 61
human-computer interaction, 7, 61

inflamation, 136
information network, 51

Jaccard's coefficient, 132

knowledge graph, 51, 150, 151
knowledge graph completion, 151
knowledge-based approach, 114
Koestler, 2

label propagation, 52
Large Language Model, 55, 161
LBD, *see* literature-based discovery
learning embeddings, 141
lemmatization, 47
link
  bridging, 3
link prediction, 52, 129, 130
LION LBD, 22
literature-based discovery, 1
LitLinker, 5
LLM, *see* Large Language Model, *see* Large Language Model
long short term memory, 54
LPA, see label propagation algorithm52
LSTM, *see* long short term memory

mean rank, 60
mean reciprocal rank, 60
measure
  distance, 49, 50
  similarity, 49
Medical Subject Headings, 5, 32
MEDLINE, 6, 31
MeSH, see Medical Subject Headings5, 32, 34, 131
MetaMap, 22
metaphor, 10
model
  ABC, 4
MR, *see* mean rank, *see* mean reciprocal rank

network, 51
network analysis, 50, 129
network embedding, 141
network node ranking, 52
neural network, 53
  CNN, 54
  deep, 53
  LSTM, 54
  transformer, 54
NN, *see* neural network

one-shot learning, 55
OntoGen, 50, 96
ontology, 36, 51
open discovery, 5, 111, 116
outlier, 93
outlier detection, 94
outlier-based discovery, 94

PageRank, 53
  Personalized, 53
part-of-speech, 47
perfect classifier, 58
Personalized PageRank, 53
plant defense, 145
POS, *see* part-of-speech
PPR, see Personalized PageRank53
PR, see PageRank53
precision, 57, 145
preface, ix
Preferential attachment, 132
prompting, 55, 162
PubMed, 31

RaJoLink, 116, 117
Random Forest, 84
random walker, 53
ranking, 86
rare term, 121
rate
  false positive, 58
  true positive, 58
Raynaud, 111
recall, 58
receiver operating characteristic, 58
relational bisociation, 143
representation
  concept, 2
reusable code, 67
reuse, 67
ROC, see receiver operating characteristic58, 84, 133
  space, 58
ROC curve, 59
  convex hull, 59
RotatE, 152

schizophrenia, 136
score propagation, 53
semantic approach, 114
semantic filtering, 114, 117
semantic profiling, 115
SemMedDB, 150
SemRep, 23, 56
sensitivity, 58
signal detection theory, 58
similarity, 49
  cosine, 49
similarity measure, 49
SimRank, 53
social network, 50
stemming, 47
stopword removal, 47
Swanson, 2
syntactic parsing, 47

tagging, 47
  part-of-speech, 47
term, 2
  bridging, 3
term frequency, 48
term frequency - inverse document frequency, 49
term ranking, 86
text embeddings, 141
text mining, 77
TextFlows, 99, 161
TF, see term frequency48
TF-IDF, see term frequency - inverse document frequency49
theory of transformative discoveries, 1
tokenization, 47
TransE, 151
transformer, 54, 55, 161
true negative, 57
true positive, 57
true positive rate, 58

UMLS, 22, see Unified Medical Language System32, 32, 114, 115
UMLS metathesaurus, 34
UMLS semantic network, 34
Unified Medical Language System, 22, 32

Weeber, 4
weighting scheme
  binary, 48
  term occurrence, 48
  TF, 48
  TF-IDF, 49
word2vec, 54, 142

www.ingramcontent.com/pod-product-compliance
Lightning Source LLC
Chambersburg PA
CBHW071656150925
32623CB00005B/432
9783031968624